Festschrift for Ned K. Johnson: Geographic Variation and Evolution In Birds

ORNITHOLOGICAL MONOGRAPHS

Editor: John Faaborg

224 Tucker Hall
Division of Biological Sciences
University of Missouri
Columbia, Missouri 65211

Project Manager: Mark C. Penrose
Managing Editor: Richard D. Earles

AOU Publications Office
622 Science Engineering
Department of Biological Sciences
University of Arkansas
Fayetteville, Arkansas 72701

The *Ornithological Monographs* series, published by the American Ornithologists' Union, has been established for major papers and presentations too long for inclusion in the Union's journal, *The Auk*. All material in this monograph may be copied for noncommercial purposes of educational or scientific advancement without need to seek permission.

Copies of *Ornithological Monographs* are available from Buteo Books, 3130 Laurel Road, Shipman, VA 22971. Price of *Ornithological Monographs* no. 63: $10.00 ($9.00 for AOU members). Add $4.00 for handling and shipping charges in U.S., and $5.00 for shipping to other countries. Make checks payable to Buteo Books.

Editors of this issue: Carla Cicero and J. V. Remsen, Jr.

Library of Congress Control Number 2007924965

Printed by Cadmus Communications, Lancaster, PA 17601

Issued 16 May 2007

*Ornithological Monographs*, No. 63 viii + 114 pp.



ISBN: 978-0-943610-75-7

Cover: "In Memory of Ned K. Johnson: Spring in Owens Valley." Original watercolor painting by Monica J. Albe.

Translation of abstracts by María José Fernández.

# Festschrift for Ned K. Johnson: Geographic Variation and Evolution In Birds

Edited by

Carla Cicero[1] and J. V. Remsen, Jr.[2]

[1]*Museum of Vertebrate Zoology, 3101 Valley Life Sciences Building, University of California, Berkeley, California 94720, USA; and*
[2]*Museum of Natural Science, Louisiana State University, Baton Rouge, Louisiana 70803, USA*



ORNITHOLOGICAL MONOGRAPHS NO. 63

PUBLISHED BY
THE AMERICAN ORNITHOLOGISTS' UNION
WASHINGTON, D.C.
2007

TABLE OF CONTENTS

# *From the Editor*

As ornithological societies go, the AOU is large, and it holds large meetings. I started to attend these as often as possible about 30 years ago, when I got a regular job and decided I wanted to be a practicing ornithologist (and I will keep practicing until I get it right). I don't know exactly when I learned who Ned Johnson was, but it seems like it must have been a long time go. Ned was large and loud and really distinctive, with a silver head of hair (I assume that at one time his hair was a different color, but I recall only that brilliant silver). Ned seemed involved in everything going on at a meeting, probably because he was. Although I never "hung out" with the taxonomy folks, Ned was outgoing enough that I, like most AOU members, considered him a friend.

*Ornithological Monograph No. 63* is dedicated to this friend of the AOU. It is the 10th of what we are calling the "new" *Ornithological Monographs*—generally smaller than the old ones and issued on a regular basis—and the second of these dedicated to a recently deceased member of the AOU. Are we setting some sort of precedent that we will have to follow every time a distinguished member and former president of the AOU dies? I don't think so. Our first memorial volume was dedicated to Ernst Mayr, whose impressive career and 100-year life span necessitated such an honor. Mayr's accomplishments included many years of service to the AOU, including a term as president. Among other things, the Mayr monograph with accompanying DVD provides an incredible look at the history of ornithology in the United States over more than half of the past century.

This tribute to Ned Johnson is different. Ned, who unfortunately didn't live as long as Ernst Mayr, was an incredible mixture of high-powered scientist and dedicated servant of scientific societies. He provided much insight into the ecology and evolution of birds, particularly those of the western United States, by adapting to new methodologies when they came along. He combined scientific collecting with molecular biology, and museum work with amazing field knowledge and experience. The following chapters include a review of Ned's major contributions, some of his last collaborative work, and an array of state-of-the-art papers on biogeography and systematics that highlight some of his interests. I am not a systematist, but I found this set of papers both interesting and a great way to update my knowledge of some contentious issues.

Ned and I got to know each other by socializing at AOU meetings, but he wasn't there just to party. Ned was president of two of the major ornithological societies (including the AOU, in 1996–1998) and contributed nearly 70 person-years of service on AOU committees, particularly the one dealing with classification. Ned's outstanding service to science, including an exemplary research career and legacy, have earned this *Festschrift* in his honor.

*John Faaborg*

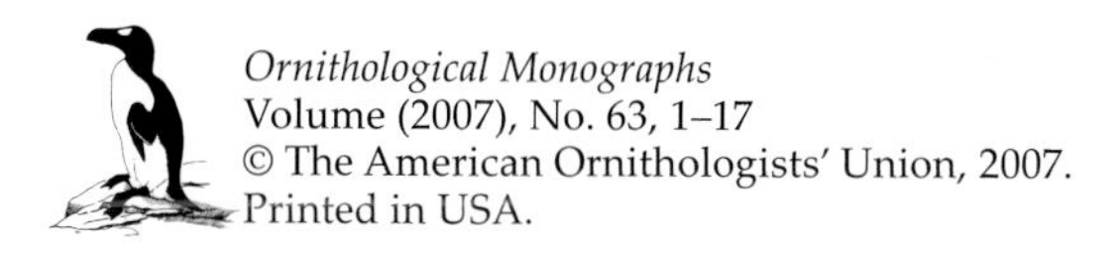
*Ornithological Monographs*
Volume (2007), No. 63, 1–17

Printed in USA.

CHAPTER 1

# A TRIBUTE TO THE CAREER OF NED K. JOHNSON: ENDURING STANDARDS THROUGH CHANGING TIMES

J. V. Remsen, Jr.[1,3] and Carla Cicero[2]

[1]*Museum of Natural Science, Louisiana State University, Baton Rouge, Louisiana 70803, USA; and* [2]*Museum of Vertebrate Zoology, University of California, Berkeley, California 94720, USA*

Abstract.—Ned K. Johnson (1932–2003) served as Curator of Birds and Professor of Zoology at the Museum of Vertebrate Zoology (MVZ), University of California, Berkeley, from 1961 to his death in 2003. His distinguished career in ornithology included major contributions on various topics (e.g., geographic variation, biogeography, speciation, systematics, integration of molt and migration) that resulted in his receiving the Brewster Medal from the American Ornithologists' Union in 1992. With eight peer-reviewed short publications in print by the time he finished his undergraduate degree at the University of Nevada, Johnson launched into his doctoral program at MVZ under Alden H. Miller. His dissertation monograph on western flycatchers (*Empidonax* spp.) provided a model of meticulous, detailed empirical analyses that represented the epitome of its genre and presaged a remarkable series of investigations on the biogeography, speciation, and geographic variation of birds. He kept pace with technological advances throughout his career, from sonograms to complex multivariate statistics to spectrophotometry and molecular genetics. Johnson's advocacy of the importance of specimen-based research has been a bulwark of strength for museum ornithologists. His contribution to MVZ of >7,200 data-rich vertebrate specimens adds an important dimension to his legacy. Johnson's consistent approach to research, namely an unimpeachable empirical foundation for addressing critical conceptual issues, continues to inspire students and colleagues to match his standards. *Received 23 January 2007, accepted 12 February 2007.*

Resumen.—Ned K. Johnson (1932–2003) fue curador de aves y profesor de zoología en el Museo de Zoología de Vertebrados (MVZ), Universidad de California, Berkeley, desde 1961 hasta su muerte en 2003. Su distinguida carrera como ornitólogo incluye importantes contribuciones en diversas áreas (variaciones geográficas, biogeografía, especiación, sistemática, relación entre muda y migración), siendo reconocido en 1992 con la medalla Brewster de la Unión de Ornitólogos Americanos. Johnson terminó sus estudios de pregrado en la universidad de Nevada con ocho publicaciones en revistas de reconocido prestigio. Inmediatamente después, comenzó el doctorado en el MVZ bajo la tutoría de Alden H. Miller. Su tesis sobre los papamoscas *Empidonax* presenta unos análisis empíricos minuciosos y detallados que constituyen un modelo en su género, presagiando una importante serie de investigaciones sobre biogeografía, especiación y variación geográfica en aves. Johnson siguió de cerca los avances tecnológicos durante su carrera, empleando desde sonogramas a estadística multivariante compleja, espectrofotometría o genética molecular. La importancia que Johnson imprimió a la investigación basada en especímenes de museo ha servido de plataforma para todos los ornitólogos que trabajan con colecciones de museo. Su contribución al MVZ, con más de 7,200 entradas de especies de vertebrados, confiere una importante dimensión a su legado. Su método de investigación, con una base empírica intachable a la hora de abordar cuestiones conceptuales críticas, continúa inspirando y sirve de ejemplo a estudiantes y colegas.

---

[3]E-mail: najames@lsu.edu

Ned K. Johnson served as Curator of Birds and Professor of Zoology at the Museum of Vertebrate Zoology (MVZ), University of California, Berkeley, from 1961 to his death in 2003. During this period, he influenced North American ornithology to a remarkable degree through his rigorous approach to specimen-based research and his emphasis on the importance of amassing detailed, accurate empirical databases for generating new ideas (see Appendix 1, and Barrowclough and Zink 2004, Cicero 2004).

## An Early Start

Johnson's research career began in high school. At the remarkably young age of 16, he published his first scientific paper in a prominent peer-reviewed journal. Reproduced here in its entirety (Fig. 1), this paper, brief as it is, nonetheless reveals some of the ingredients of his distinguished career. First, note that he already was collecting birds for the University of Nevada Museum in Reno. Second, it is clear that he already was attuned to bird behavior and its significance. Third, he had the precocial wherewithal not only to recognize the significance of his observation but also to publish it in one of the best ornithological journals, all this in 1949, from the relative isolation of western Nevada.

By the time he was 21, Johnson had published seven additional short papers in *The Condor, Journal of Mammalogy,* and *Great Basin Naturalist* (Appendix 1) on birds and mammals of Nevada and had graduated with a B.S. from the University of Nevada, Reno (UNR). Instrumental to his early success was his mentor Ira LaRivers, a professor at UNR and an expert on the insects and fishes of Nevada. Although not an ornithologist, LaRivers encouraged Johnson to collect birds by adding him as a subpermittee on his collecting permit and taking him on numerous field trips throughout the state. Early on, Johnson realized the importance of scientific collecting for careful documentation of avian distribution and variation.

After two years in the U.S. Army in Germany, Johnson entered the doctoral program in the Department of Zoology, University of California, Berkeley, with Alden H. Miller at the MVZ as his advisor. At this time, after many decades of visionary leadership by Joseph Grinnell and Miller (Johnson 1995a), the MVZ was almost without rival as the world's leading university-based research museum, and the Grinnell-Miller tradition of specimen-based research was clearly what drew Johnson to the MVZ.

## Geographic Variation and Speciation

Johnson finished his Ph.D. in 1961, and publication of his dissertation soon followed (Johnson 1963a). Following decades of MVZ tradition, the dissertation was published in the *University of California Publications in Zoology* (UCPZ) series. This monograph was significant in many ways. Although Johnson's analyses included voluminous state-of-the-times morphometric analyses of exceptionally large series of specimens (3,253 in this case) that typified MVZ monographs in the UCPZ series, Johnson's was the first of the genre to include major analyses of vocalizations using sonograms. Johnson's monograph introduced quantitative analyses of the fine details of vocalizations as a major axis in the study of geographic variation. It thus contained a trademark of Johnson's career that carried through to his final years: adoption of the latest methods to address relevant questions. Also established in this monograph was Johnson's interest in "difficult" groups. Prior to his research, the *Empidonax* flycatchers of western North America were the least-understood group on the continent, with

**Loggerhead Shrike Steals Shot Sparrow.**—In early May, 1948, my brother and I were collecting birds at a small marsny pond in a grassy field about four and one-half miles southeast of Reno, Nevada. A Savannah Sparrow (*Passerculus sandwichensis*) flew up to a fence post about forty feet from where we stood. My brother snot the bird. The second it hit the ground, we ran to pick it up but were amazed to see a Loggerhead Shrike (*Lanius ludovicianus*) seize our bird and carry it away. The shrike carried the sparrow with laborious strokes of its wings and flew to a spot in a large field of sagebrush about fifty yards away. Because the prey was comparatively heavy, the shrike flew low over the ground. It seems incredible but we could not locate either the shrike or the sparrow after searching the area for at least a half hour.—Ned K. Johnson, *Reno, Nevada, February 13, 1949.*

Fig. 1. Ned K. Johnson's first publication, in *The Condor,* at age 16 (Johnson 1949; see Acknowledgments).

nagging questions of how many species really existed and how they differed from one another. Johnson discovered how they could be identified in all age, sex, and seasonal categories and how they differed ecologically and vocally. His measurement data are impressive in their detail (e.g., individual measurements of the six outermost primaries for each of three age categories for each sex of each species).

Johnson's interest in *Empidonax* flycatchers as models for the study of intraspecific geographic variation and of differences between closely related, extraordinarily similar species continued throughout his career. Although most studies of intraspecific variation gravitate toward species that show marked geographic variation, Johnson followed his UCPZ monograph with a study of a species of *Empidonax*, Hammond's Flycatcher (*E. hammondii*), that shows virtually no geographic variation despite a breeding distribution that extends from Alaska to the southwestern United States. Such an extensive latitudinal distribution typically guarantees at least some strong geographic variation in size-related morphology, but Johnson (1966:198) demonstrated that the species showed essentially no geographic variation in morphometrics or plumage:

> The analysis is based on 545 specimens of summer residents which were divided for treatment into 13 geographic populations, were segregated by sex and age, and were measured for the length of the tenth primary, tail length, bill length, bill depth, bill width, tarsus length, middle toe length, and body weight.... This species is remarkably uniform throughout its range....

Johnson pointed out the potential significance of this seemingly "negative" finding: "Reduction or absence of geographic variation...seems best attributed to strong genetic homeostasis" (Johnson 1966:199).

In 1980, he published a second UCPZ monograph on *Empidonax* (Johnson 1980), this time focusing on the *difficilis–flavescens* group. In this monograph, Johnson added mulitvariate statistics and spectrophotometry to his repertoire of techniques; the spectrophotometry represented one of the first applications of this technique to analysis of geographic variation in bird coloration. As was typical of Johnson's papers, he included a theoretical section that presented a model for speciation mechanisms and diversification in the genus, with applications and implications far beyond *Empidonax* flycatchers. Unfortunately, both of Johnson's monographs on *Empidonax* are under-read and under-cited in bird biology. The following are samples of some of the significant findings from these two monographs.

> The important point to be stressed here is that each of the species *hammondii, oberholseri, wrightii, affinis,* and *minimus*...has a characteristic wing shape, which, although varying somewhat within each form, can be an extremely useful criterion for identification when interpreted in the light of variation with sex and age. (Johnson 1963a:94)

> [T]he habitat preference of each species is a characteristic trait which probably evolved far back during the geographic isolation and speciation of each form at the same time that morphologic and behavioral divergence was taking place. (Johnson 1963a:189)

> The advertising songs, distinctive in each of the three species, as proved by the physical analysis of tape recordings with a sonagraph, function chiefly in pair-bond establishment and maintenance; their role in territorial defense is relatively minor.... *oberholseri* and *wrightii* maintain territories both intra- and interspecifically by means of hostile displays and vocalizations. (Johnson 1963a:215)

> The broad continental pattern, of alternation between long-tailed with short-tailed groups of populations as one moves from the Channel Islands to the coast, through the interior, down through Mexico and into Central America, is again seen. (Johnson 1980:46)

> Purity of breast color reveals a beautifully clinal geographic variation. Three clear steps exist in this cline, a slight one between Coahuila and San Luis Potosí, a moderate one at the Isthmus of Tehuantepec, and a strong one between Nuclear Central America and Costa Rica. (Johnson 1980:56)

> The male position note provides an exceptionally clear example of geographic trends in a simple behavioral signal. Broad regions of uniformity in note structure are separated by relatively narrow belts where the character changes abruptly or where a mixture of note types is seen. (Johnson 1980:70)

> Whole complexes of features change more or less simultaneously over fairly narrow and

> well-defined zones. Between these zones are broad regions where character expression is comparatively uniform. (Johnson 1980:102)

With the advent of biochemical techniques to measure genetic variation, Johnson, in collaboration with then lab technician Jill Marten, was one of the leaders in applying them to the study of geographic variation. This was clearly an exciting time for Johnson; he could now explore the relationship between phenotypic and genotypic variation. Protein electrophoresis in the 1980s and DNA sequencing in the 1990s became standard techniques in Johnson's studies of geographic variation, allowing new perspectives on his studies of *Empidonax* (e.g., Johnson and Marten 1985, 1988, 1991; Johnson and Cicero 2002) as well as forming a framework for initiation of studies on sparrows (*Amphispiza* spp.; Johnson and Cicero 1991; Johnson and Marten 1992; Cicero and Johnson 2006, 2007) and vireos (*Vireo* spp.; Cicero and Johnson 1992, Johnson 1995b). Johnson, with his doctoral student Bob Zink and postdoctoral researcher George Barrowclough as collaborators, was at the forefront of the application of genetic techniques to studies of speciation patterns, not only in *Empidonax* (e.g., Zink and Johnson 1984) but also in sapsuckers (*Sphyrapicus* spp.; Johnson and Zink 1983) and vireos (e.g., Johnson and Zink 1985, Johnson et al. 1988). In addition, they also formulated early reviews and conceptual models of genetic diversification (Barrowclough et al. 1984, Barrowclough and Johnson 1988). Johnson later revisited these groups, using newer techniques (DNA sequencing) and his unequaled knowledge of these birds to map ecological and morphological characters onto the phylogeny to explore patterns of diversification (Cicero and Johnson 2002). The following are samples of the major conclusions of these analyses.

On *Empidonax*:

> The pattern resulting from evolution within the genus *Empidonax* resembles more of a "bush" than a dichotomous sequence of cladogenic events.... We conclude, therefore, that external morphological similarity among species is due to developmental canalization of an *Empidonax* phenotype, rather than to an extremely recent origin of most species. (Zink and Johnson 1984: 213–214)

> Occupancy of temperate habitats by certain genera in this group is coincident with their evolution of migratory behavior and with independent diversification in foraging modes that reduces potential competition in sympatry. (Cicero and Johnson 2002:289)

> A major result of this study is the strong congruence between mtDNA sequences and other characters (allozymes, morphology, behavior).... (Cicero and Johnson 2002:297)

On sapsuckers:

> Therefore, the finding of genic divergence and phenotypic similarity (mosaic evolution) suggest that *varius* and *nuchalis* have simply retained the ancestral plumage condition, while evolution has proceeded at the genic level, probably in a more-or-less uniform, time-dependent manner. (Johnson and Zink 1983:880)

> Our mtDNA data firmly establish the conclusions presented by Johnson and Zink (1983) concerning evolutionary relationships among North American sapsuckers. In particular, the nearly perfect association between genetic distances estimated by allozymes and mtDNA sequences provides incontrovertible evidence that *ruber* and *nuchalis* shared a very recent common ancestor.... (Cicero and Johnson 1995:560)

On vireos:

> We speculate that "*chivi*" arose from wintering individuals of *V. olivaceus* that failed to return to North America. (Johnson and Zink 1985:433)

> Molecular phylogenetic evidence suggests that migration and seasonal habitat shifts evolved multiple times in the radiation of vireos. (Cicero and Johnson 1998:1367)

In investigating the mechanics of the speciation process, Johnson assembled an impressive data set on mate preferences in the contact zone between *Sphyrapicus ruber* and *S. nuchalis* (Johnson and Johnson 1985). This study, which is the primary data set supporting the ranking of the component taxa as separate biological species (Red-breasted and Red-naped sapsuckers; American Ornithologists' Union [AOU] 1998), consisted of 145 mated pairs in which both members were collected so that phenotype and genotype could be assessed. Assessments of mate preference and extent of hybridization led to the following conclusions:

> [A]lthough parental phenotypes predominate in the hybrid zone, $F_1$ hybrids seem to enjoy equivalent viability, and their occurrence is in proportion to the frequency of interspecific matings.... $F_2$ hybrids, however, do seem to be at a disadvantage. (Johnson and Johnson 1985:11)

> It is well established that *daggetti* arrives on the sympatric nesting areas ahead of *nuchalis*.... We speculate that the reddest males of *daggetti* dominate the less-red males, win the best territories quickly, and are the earliest to gain mates.... Some of these paler male *daggetti* possess territories before any *nuchalis* males. A small percentage of female *nuchalis* choose established male *daggetti* as mates because of the superstimulus value of the extensive red. (Johnson and Johnson 1985:13)

As a follow-up to the morphological study of *E. hammondii*, Johnson and Marten (1991) used allozymes to compare levels of genetic variation with the striking morphological homogeneity observed across the broad latitudinal breeding distribution of this species. At the time of this study, "genetic analyses of breeding samples of birds across most of their range [were] especially rare" (Johnson and Marten 1991:232). The genetic data paralleled those for morphology, with minimal population structuring, low levels of variability, and high estimates of gene flow. In interpreting these findings, Johnson and Marten (1991:236–237) concluded that

> The essentially homogeneous genetic structure of the species across its broad modern range agrees with a scenario of moderately rapid expansion into newly available habitat from a previously confined distribution.... [T]he ancestral stock could have moved into the Pacific Northwest region during the early Holocene from a Boreal refuge of forest or woodland located anywhere south of Cordilleran ice.... The absence of geographic variation in morphology could also be a consequence of reduced genetic variability, a suggestion that assumes that low genetic variability at enzyme-coding loci reflects reduced variability at loci controlling the expression of the external phenotype.... [I]f the postglacial distribution is as recent as implied above, it is plausible to suggest that insufficient time has elapsed for the evolution of observable phenotypic variation.

This early connection between glaciation and genetic and morphological patterns in birds continues to shape current thinking about processes of speciation and biogeography (e.g., Johnson and Cicero 2004, Lovette 2005).

Johnson's studies of geographic variation and speciation were based on strong and outspoken support for a Biological Species Concept that allows for limited hybridization (see Johnson et al. 1999, Winker et al. 2007), and he believed that "essential genetic independence resulting from reproductive isolation...is responsible for the evolution of avian biodiversity" (Johnson et al. 1999:1470). He also believed that speciation can occur relatively rapidly and that genetic divergence alone is insufficient to determine species status (Johnson and Cicero 2004). By applying a variety of field- and museum-based approaches to understanding species limits in birds (e.g., morphology, genetics, vocalizations, distribution, ecology), he produced definitive papers that often took years to complete but encompass the suite of traits important for reproductive isolation.

In sum, Johnson's approach to the study of geographic variation and speciation was to (1) accumulate large samples of target species from key areas through extensive field work, with a focus on breeding material for migratory species and fresh fall-plumaged material for resident species, (2) study the specimens in meticulous detail using traditional techniques as well as the latest technology, (3) analyze those empirical data using advanced analytical techniques, and (4) synthesize the results and evaluate their generality and theoretical significance. The outcome of this system then was used to plan the next season's field work, and this "annual cycle" was repeated unfailingly until a few months before Johnson's death. The patience and zeal with which Johnson pursued these studies is a hallmark of his work.

### Molt–Migration–Life History

Stimulated by the marked differences among superficially similar species of *Empidonax*, Johnson was a leader in the study of integrating molt and migration into the comparison of life-history patterns among closely related species. These studies, done relatively early in his career, included a comparison of molt cycles in *Empidonax* (Johnson 1963b) and analyses of molt or migratory patterns in *E. hammondii* (Johnson 1963a, 1965, 1970), *E. oberholseri* and *E. wrightii* (Johnson 1963a), *E. difficilis* (Johnson 1973,

1974a), and *E. flavescens* (Johnson 1974a). At the time of his death, Johnson also was comparing tail molt patterns among tyrannid flycatchers (unpublished). As with his morphological data, specimens were divided by sex and age classes in addition to geography. Johnson (1963b:879) found that the

> most obvious relationship between molt and migration in *Empidonax* is that molts occurring after migration tend to be protracted. In other words, an early molt on the breeding grounds is correlated with a leisurely, prolonged southward migration, and an early and rapid fall migration is associated with a subsequent protracted molt.

Other important findings were that populations of *E. hammondii* differed geographically and seasonally in their migration patterns, and that juvenile mortality was high during fall migration (Johnson 1970:182–185):

> In the spring migration of this species, an early rapid migration in coastal regions contrasts with a protracted movement northward through the interior.... An opposite pattern is shown by the fall movement, in which there is a leisurely migration through coastal areas compared to an early and comparatively rapid passage southward from the interior.... Differences in age composition between autumn and winter samples point to a comparatively high mortality of immatures in the late fall and early winter.

Both for individual *Empidonax* species and for the genus as a whole, Johnson concluded that such differences in molt and migration patterns are adaptive: "In my opinion, these differences between coastal and interior migrants [of *E. hammondii*] are adaptive..." (Johnson 1970:182) and "the degree of interspecific variation in timing is so great as to suggest that the exact position in the annual cycle of these molts is chiefly a reflection of adaptation at the species level" (Johnson 1963b:883).

## Pigmentation and Coloration

Johnson was fascinated with pigmentation and the evolution of color differences in birds and published several significant papers on this topic that probably have not received the attention they deserve. Although he routinely used color spectrophotometry to study patterns of color variation within and among populations (e.g., Johnson 1980, Johnson et al. 1998), he wanted to go beyond that technique to understand the pigmentary basis for such variation. Thus, he collaborated with Alan Brush to apply pigment chemistry in addition to colorimetry in studies of tanagers (*Chlorospingus* spp.; Johnson and Brush 1972) and warblers (*Vermivora* spp.; Brush and Johnson 1976). The study of the tanagers involved *C. pileatus* and *C. "zeledoni,"* which at the time were considered separate taxa that exhibited local sympatry on two high mountains in central Costa Rica. Using these methods, Johnson and Brush (1972) elegantly corroborated prior suggestions that the two taxa actually represent different color phases of the same species. By extracting pigment from the breast plumage of a single specimen (MVZ 162203), they transformed it from the "gray-green morph" to the "yellow-green morph." In a major finding, they proposed

> that on Volcanes Irazú and Turrialba genotypes responsible for increased grayness (= reduced yellowness), operating through decreased concentration of lutein pigmentation, owe their survival and maintenance to the long history of vulcanism that has characterized the central highlands. (Johnson and Brush (1972:260)

Johnson also integrated his interest in color and plumage pigmentation with his interest in molt, natural history, and the annual cycle. In a study of the Green Jay (*Cyanocorax yncas longirostris*) in seasonally dry deciduous woodland in the mid-Marañon Valley of Peru (Johnson and Jones 1993), he divided specimens by collection date as well as age and analyzed them for spectral reflectance, molt, and feather wear. Unlike other subspecies that occupy more humid habitats and are dorsally green year-round, individuals in this population gradually fade from bright yellow-green to greenish-blue or blue dorsally. This study showed that "autoxidation and accompanying bleaching from exposure to sunlight [as opposed to feather wear] are implicated in this striking color change" (Johnson and Jones 1993:389).

More recently, Johnson dove into the study of pigmentary properties of feathers. As a mentor for undergraduates, he worked with a student examining the basis for color variation in the "red" crown plumage of woodpeckers (Picidae;

Hopenstand and Johnson 2001). In another paper (Johnson 2001), he presented a theory for the evolutionary origin of feathers:

> I propose that the "aerodynamic features" of modern feathers, including a distinctive combination of flatness, light weight, resilience, and smooth-contour, initially evolved in a context unrelated to flight. Instead, flatness and lightness evolved in response to selection favoring maximally effective shape for the exhibition of pigments and structural colors.... The physical characteristics and individual mobility of feathers, serving to expose pigments and iridescece during display, were preadaptive for the eventual use of feathers in flight and thermoregulation. (Johnson 2001:91)

### Biogeography and Natural Range Expansions

Johnson's voluminous contributions to the documentation of breeding distributions of insular montane avifaunas in the western United States naturally provoked his interest in the patterns and their underlying processes. Aware that distributions can change with lightning speed compared with geological time, he focused primarily on ecological processes that might govern the patterns. One of his major contributions to the field (Johnson 1975), conducted when studies of island biogeography were "cutting edge," showed that almost all of the pattern on montane islands could be explained not by distance and area effects, much less historical biogeography, but by the simplest of ecological variables, namely habitat availability:

> Total area and area of forest-woodland, variables of significance in other, similar studies, here predict only from 28 to 45% of the variation in total species number. In contrast, an index of habitat diversity explains 91% of the variation in total bird species. (Johnson 1975:561)

In another important contribution, Johnson (1978) showed that avifaunal distributions, contact zones, and areas of active speciation in the western Intermountain Region are coincident with physiographic, climatic, and floristic discontinuities.

Although Johnson was interested primarily in continental biogeography, especially in the Great Basin, he also published two important papers on the California Channel Islands (Johnson 1972, Lynch and Johnson 1974). In the first, he reviewed the composition of the avifauna on different islands and considered evidence explaining the distribution and differentiation (or lack thereof) of island populations. The second paper involved a literature survey of minimum avifaunal turnover rates (immigration vs. extinction) on islands, including those reported by Diamond (1969) for the Channel Islands. Diamond (1969) reviewed early distributional records summarized by Howell (1917), compared those with his own records, and concluded that although the total number of resident bird species on each island has remained fairly constant over time, the species composition of the avifauna has changed markedly on most islands, which suggests high turnover (17–62%). In a bold critique, Lynch and Johnson (1974) faulted Diamond for not publishing the majority of records and species' identities on which his rates were based, thus making his figures "impossible to interpret properly" (Lynch and Johnson 1974:375). They also noted that

> the great majority of extinctions reported by Diamond to have occurred on the California Channel Islands are attributable either to human interference or to faulty interpretation of faunal data (pseudoturnover). While a few "natural" extinctions may have occurred, we can find only a single reasonably well-documented example out of 41 specified extinctions. (Lynch and Johnson 1974:378)

Therefore, on the basis of their literature survey, they concluded that minimum avifaunal turnover rates on islands have been seriously overestimated, especially in regard to natural processes.

Johnson's interest in biogeography and his intensive field work through the decades in the western United States put him in a unique position to detect and document avifaunal range changes (Johnson 1974b, 1994, 1995c; Johnson and Garrett 1974, Johnson and Cicero 1985). The changes in which he was interested did not involve obvious responses to modification of the landscape by humans; rather, he focused on regions and habitats with minimal direct human influence, such as coniferous forests on remote mountain ranges. The significance of this type of range change is clearly much greater, as Johnson (1994) noted years in advance of widespread attention to global climate change:

> These range adjustments are not responses to anthropogenic influences. Instead, climatic change in the new regions of occupancy apparently has provided regimes of increased summer moisture and higher mean summer temperature typical of pre-expansion distributions. (Johnson 1994:27)

His interest in these phenomena culminated in a landmark symposium volume on avifaunal change in western North America, co-edited with J. R. Jehl (Jehl and Johnson 1994). In keeping with his interest in applying new methods to ongoing problems, Johnson would have been excited to see the application of climatic and ecological niche modeling in detailing natural range changes in birds (e.g., Barred Owl [*Strix varia*]; Monahan and Hijmans 2007).

### Reviews

Johnson published what we suspect is an exceptional number of book, symposium, and monograph reviews. He was the unquestioned authority on birds of the Great Basin and geographic variation of birds in western North America, as well as being highly respected for his general knowledge of bird biology. Thus, editors turned to Johnson for many important reviews, such as of the *Birds of North America* series and the *Handbook of the Birds of the World*. Johnson's reviews were not only fair but also always insightful. Our personal favorite is one of his earliest, of Lester Short's research on flickers (Johnson 1969). In this review, he proposed a novel, alternative hypothesis for the phenotypic patterns found by Short, one that represented the equivalent of a null hypothesis for genetic variation:

> The philosophy underlying Short's entire discussion is that traces in one population, A, of characters expressed in another population, B, and presumably evolved when A and B were not in contact, means that genes from population A are infiltrating population B because of secondary contact.... However appealing this explanation may be for certain situations of hybridization in birds, for the North American flickers I feel that Short's interpretation is incorrect.... The likely explanation of most of the occurrence of red nuchal traces, then, in *most* of the populations of *cafer* in western North America away from the hybrid zones, is that these traces have their genetic basis deep in the stock that gave rise to all flickers and their relatives.... (Johnson 1969: 227–228)

This talent for original ideas was also a trademark of Johnson's in public seminars and meetings. Whether the topic was close to his research or remarkably distant, Johnson routinely posed insightful questions to speakers that revealed how closely he followed their presentations.

### Specimen Contributions

Johnson's specimen catalog contained entries up to number 7,212. These were primarily birds, with some mammals, reptiles, and amphibians, most of which are deposited at the MVZ; some of his earlier material is deposited in the Museum of Biology, University of Nevada, Reno. This number of specimens is probably higher than those collected by his peers in academic ornithology and represents a lasting contribution and legacy. The beautiful quality of his specimens, the geographic importance of the specimens collected, and the quantity and quality of data on their labels, overshadows in value the sheer volume of specimens. With at least 10–12 data fields per label, Johnson directly contributed roughly 72,120–86,544 "data cells" to vertebrate biology, which would certainly rank him among the most important contributors ever among tenure-track ornithologists. Further, this does not take into account the many thousands of research specimens contributed by his students and associates.

Many ornithologists who conduct specimen-based research directly contribute little to these resources, the quality and volume of which are often the rate-limiting step in such research. Johnson was the antithesis of this approach. Not only was his field work collection-oriented, but he also believed philosophically that direct involvement generated original insights and better research. Johnson respected colleagues with a similar philosophy and disdained those whose specimen contributions were minuscule, particularly if they were employed as curators of research collections.

### Professional Service

Johnson strongly believed in service to professional societies and to the University of

California, Berkeley, where he worked for 42 years. Although outwardly cynical about anything having to do with authority and bureaucracy, he felt that it was his duty to serve in positions of influence. Thus, he was President of both the Cooper Ornithological Society (1981–1983) and the AOU (1996–1998), and was involved in many other professional organizations, including the International Ornithological Committee. His 69 committee-years of service to the AOU, including 36 years on the Committee on Classification and Nomenclature, led to his winning the Marion Jenkinson Service Award from the AOU in 2001. At UC Berkeley, Johnson helped to define the undergraduate curriculum in biology and fought for the importance of field-based organismal courses in a growing era of molecular and cell biology. In addition to serving as a mentor or advisor for countless undergraduate students, Johnson also sponsored 3 Master's and 15 Ph.D students (Appendix 2). Arguably his greatest professional legacy was the enthusiasm for natural history that he instilled in students at all levels—students who, years later, still remember Johnson's infectious passion while teaching Ornithology and the wildly popular class on Natural History of the Vertebrates.

## Scientific Standards

In the tradition of his MVZ predecessors, Grinnell and Miller, Johnson was a strong advocate of accuracy and rigor in all aspects of data gathering. With missionary zeal, he inspired his students and colleagues to strive for the highest standards of authenticity and repeatability, and always insisted on meticulous detail in data collection—from the label on a specimen, which he considered sacrosanct, to the analysis and interpretation of data. Johnson was noted for his candor in both the spoken and the written word, and often spoke up for what he believed was true, regardless of the consequences. To that end, he published an influential paper on standards for scientific specimens (Johnson et al. 1984) and also took to task (Lynch and Johnson 1974; Johnson 1994; Cicero and Johnson 1996, 2006) those whose empirical database was of poor quality. A quotation from Lynch and Johnson (1974) captures this philosophy:

> Indeed, a respected colleague has offered a friendly admonition to the effect that our standards for acceptable faunal data are so stringent that, if generally adopted, they would prevent a lot of good research from being done. We are compelled to reply that we are unaware of any examples of good research in the area of faunal analysis which do *not* involve attention to detail and a realization of the inherent limitations of observational data (Lynch and Johnson 1974:383).

We like to imagine that as Ned died, he still dreamed of collecting sparrows in the Great Basin, while a rival shrike, with silvery sagebrush-matching sheen comparable to his own, races him for the specimen.

## Acknowledgments

We are grateful to G. F. Barrowclough and J. R. Jehl, Jr., for excellent suggestions for improving the manuscript. We thank D. Dobkin for providing permission to reprint material held in copyright by the Cooper Ornithological Society (Fig. 1).

## Literature Cited

American Ornithologists' Union. 1998. Check-list of North American Birds, 7th ed. American Ornithologists' Union, Washington, D.C.

Barrowclough, G. F., and N. K. Johnson. 1988. Genetic structure of North American birds. Pages 1630–1638 and 1669–1673 *in* Acta XIX Congressus Internationalis Ornithologici (H. Ouellet, Ed.). National Museum of Natural Sciences, University of Ottawa Press, Ottawa.

Barrowclough, G. F., N. K. Johnson, and R. M. Zink. 1984. On the nature of genic variation in birds. Pages 135–154 *in* Current Ornithology, vol. 2 (R. F. Johnston, Ed.). Plenum Press, New York.

Barrowclough, G. F., and R. M. Zink. 2004. Obituary: Ned K. Johnson, 1932–2003. Ibis 146: 567–568.

Brush, A. H., and N. K. Johnson. 1976. The evolution of color differences between Nashville and Virginia's warblers. Condor 78:412–414.

Cicero, C. 2004. In Memoriam: Ned Keith Johnson, 1932–2003. Auk 121:600–602.

Cicero, C., and N. K. Johnson. 1992. Genetic differentiation between populations of Hutton's Vireo (Aves: Vireonidae) in disjunct allopatry. Southwestern Naturalist 37:344–348.

Cicero, C., and N. K. Johnson. 1995. Speciation in sapsuckers (*Sphyrapicus*): III. Mitochondrial-DNA sequence divergence at the cytochrome-*b* locus. Auk 112:547–563.

Cicero, C., and N. K. Johnson. 1996. Declining scientific standards in studies of avian distribution. [Review of *California Birds: Their Status and Distribution* by Arnold Small.] Auk 113:522–523.

Cicero, C., and N. K. Johnson. 1998. Molecular phylogeny and ecological diversification in a clade of New World songbirds (genus *Vireo*). Molecular Ecology 7:1359–1370.

Cicero, C., and N. K. Johnson. 2002. Phylogeny and character evolution in the *Empidonax* group of tyrant flycatchers (Aves: Tyrannidae): A test of W. E. Lanyon's hypothesis using mtDNA sequences. Molecular Phylogenetics and Evolution 22:289–302.

Cicero, C., and N. K. Johnson. 2006. Diagnosability of subspecies: Lessons from Sage Sparrows (*Amphispiza belli*) for analysis of geographic variation in birds. Auk 123:266–274.

Cicero, C., and N. K. Johnson. 2007. Narrow contact of desert sage sparrows (*Amphispiza belli nevadensis* and *A. b. canescens*) in the Owens Valley, eastern California: Evidence from mitochondrial DNA, morphology, and GIS-based niche models. Pages 78–95 *in* Festschrift for Ned K. Johnson: Geographic Variation and Evolution In Birds (C. Cicero and J. V. Remsen, Jr., Eds.). Ornithological Monographs, no. 63.

Diamond, J. 1969. Avifaunal equilibria and species turnover rates on the Channel Islands of California. Proceedings of the National Academy of Sciences USA 64:57–63.

Hopenstand, D., and N. K. Johnson. 2001. Variation in pigmentary and colorimetric properties of red crown feathers in selected woodpeckers (Aves: Picidae). Berkeley Scientific 5:56–60.

Howell, A. B. 1917. Birds of the islands off the coast of southern California. Pacific Coast Avifauna, no. 12.

Jehl, J. R., and N. K. Johnson, Eds. 1994. A century of avifaunal change in western North America. Studies in Avian Biology, no. 15.

Johnson, N. K. 1949. Loggerhead Shrike steals shot sparrow. Condor 51:233.

Johnson, N. K. 1963a. Biosystematics of sibling species of flycatchers in the *Empidonax hammondii–oberholseri–wrightii* complex. University of California Publications in Zoology 66:79–238.

Johnson, N. K. 1963b. Comparative molt cycles in the tyrannid genus *Empidonax*. Pages 870–883 *in* Proceedings XIII International Ornithological Congress (C. G. Sibley, Ed.). American Ornithologists' Union, Washington, D.C.

Johnson, N. K. 1965. Differential timing and routes of the spring migration in the Hammond Flycatcher. Condor 67:423–437.

Johnson, N. K. 1966. Morphologic stability versus adaptive variation in the Hammond Flycatcher. Auk 83:179–200.

Johnson, N. K. 1969. Review: Three papers on variation in flickers (*Colaptes*) by Lester J. Short, Jr. Wilson Bulletin 81:225–230.

Johnson, N. K. 1970. Fall migration and winter distribution of the Hammond Flycatcher. Bird-Banding 41:169–190.

Johnson, N. K. 1972. Origin and differentiation of the avifauna of the Channel Islands, California. Condor 74:295–315.

Johnson, N. K. 1973. Spring migration of the Western Flycatcher, with notes on seasonal changes in sex and age ratios. Bird-Banding 44:205–220.

Johnson, N. K. 1974a. Molt and age determination in Western and Yellowish flycatchers. Auk 91: 111–131.

Johnson, N. K. 1974b. Montane avifaunas of southern Nevada: Historical change in species composition. Condor 76:334–337.

Johnson, N. K. 1975. Controls of number of bird species on montane islands in the Great Basin. Evolution 29:545–567.

Johnson, N. K. 1978. Patterns of avian geography and speciation in the Intermountain Region. Great Basin Naturalist Memoirs 2:137–159.

Johnson, N. K. 1980. Character variation and evolution of sibling species in the *Empidonax difficilis–flavescens* complex (Aves: Tyrannidae). University of California Publications in Zoology, no. 112.

Johnson, N. K. 1994. Pioneering and natural expansion of breeding distributions in western North American birds. Pages 27–44 *in* A Century of Avifaunal Change in Western North America (J. R. Jehl, Jr., and N. K. Johnson, Eds.). Studies in Avian Biology, no. 15.

Johnson, N. K. 1995a. Ornithology at the Museum of Vertebrate Zoology, University of California, Berkeley. Pages 183–221 *in* Contributions to the History of North American Ornithology (W. E. Davis, Jr., and J. A. Jackson, Eds.). Memoirs of the Nuttall Ornithological Club, no. 12.

Johnson, N. K. 1995b. Speciation in vireos. I. Macrogeographic patterns of allozymic variation in the *Vireo solitarius* complex in the contiguous United States. Condor 97:903–919.

Johnson, N. K. 1995c. Seven avifaunal censuses spanning one-half century on an island of white firs (*Abies concolor*) in the Mojave Desert. Southwestern Naturalist 40:76–85.

Johnson, N. K. 2001. The role of pigments and structural colors in the evolutionary origin of feathers. Evolutionary Theory 12:91–97.

Johnson, N. K., and A. H. Brush. 1972. Analysis of polymorphism in the Sooty-capped Bush Tanager. Systematic Zoology 21:245–262.

Johnson, N. K., and C. Cicero. 1985. The breeding avifauna of San Benito Mountain, California:

Evidence for change over one-half century. Western Birds 16:1–23.

Johnson, N. K., and C. Cicero. 1991. Mitochondrial DNA sequence variability in two species of sparrows of the genus *Amphispiza.* Pages 600–610 *in* Acta XX Congressus Internationalis Ornithologici (B. D. Bell, Ed.). Congress Trust Board, Wellington, New Zealand.

Johnson, N. K., and C. Cicero. 2002. The role of ecologic diversification in sibling speciation of *Empidonax* flycatchers (Tyrannidae): Multigene evidence from mtDNA. Molecular Ecology 11: 2065–2081.

Johnson, N. K., and C. Cicero. 2004. New mitochondrial DNA data affirm the importance of Pleistocene speciation in North American birds. Evolution 58:1122–1130.

Johnson, N. K., and K. L. Garrett. 1974. Interior bird species expand breeding ranges into southern California. Western Birds 5:45–56.

Johnson, N. K., and C. B. Johnson. 1985. Speciation in sapsuckers (*Sphyrapicus*): II. Sympatry, hybridization, and mate preference in *S. ruber daggetti* and *S. nuchalis*. Auk 102:1–15.

Johnson, N. K., and R. E. Jones. 1993. The Green Jay turns blue in Peru: Interrelated aspects of the annual cycle in the arid tropical zone. Wilson Bulletin 105:388–398.

Johnson, N. K., and J. A. Marten. 1985. The geography of allelic frequencies in the Western Flycatcher (*Empidonax difficilis*). Pages 1029–1030 *in* Proceedings XVIII International Ornithological Congress (V. D. Ilyichev and V. M. Gavrilov, Eds.). Nauka Publishers, Moscow.

Johnson, N. K., and J. A. Marten. 1988. Evolutionary genetics of flycatchers. II. Differentiation in the *Empidonax difficilis* complex. Auk 105:177–191.

Johnson, N. K., and J. A. Marten. 1991. Evolutionary genetics of flycatchers. III. Variation in *Empidonax hammondii* (Aves: Tyrannidae). Canadian Journal of Zoology 69:232–238.

Johnson, N. K., and J. A. Marten. 1992. Macrogeographic patterns of morphometric and genetic variation in the Sage Sparrow complex. Condor 94:1–19.

Johnson, N. K., J. V. Remsen, Jr., and C. Cicero. 1998. Refined colorimetry validates endangered subspecies of the Least Tern. Condor 100:18–26.

Johnson, N. K., J. V. Remsen, Jr., and C. Cicero. 1999. Resolution of the debate over species concepts in ornithology: A new comprehensive biologic species concept. Pages 1470–1482 *in* Acta XXII Congressus Internationalis Ornithologici (N. J. Adams and R. H. Slotow, Eds.). BirdLife South Africa, Johannesburg.

Johnson, N. K., and R. M. Zink. 1983. Speciation in sapsuckers (*Sphyrapicus*): I. Genetic differentiation. Auk 100:871–884.

Johnson, N. K., and R. M. Zink. 1985. Genetic evidence for relationships among the Red-eyed, Yellow-green, and Chivi vireos. Wilson Bulletin 97:421–435.

Johnson, N. K., R. M. Zink, G. F. Barrowclough, and J. A. Marten. 1984. Suggested techniques for modern avian systematics. Wilson Bulletin 96:543–560.

Johnson, N. K., R. M. Zink, and J. A. Marten. 1988. Genetic evidence for relationships in the avian family Vireonidae. Condor 90:428–445.

Lovette, I. 2005. Glacial cycles and the tempo of avian speciation. Trends in Ecology and Evolution 20:57–59.

Lynch, J. F., and N. K. Johnson. 1974. Turnover and equilibria in insular avifaunas, with special reference to the California Channel Islands. Condor 76:370–384.

Monahan, W. B., and R. J. Hijmans. 2007. Distributional dynamics of invasion and hybridization by *Strix* spp. in western North America. Pages 55–66 *in* Festschrift for Ned K. Johnson: Geographic Variation and Evolution in Birds (C. Cicero and J. V. Remsen, Jr., Eds.). Ornithological Monographs, no. 63.

Winker, K., D. A. Rocque, T. M. Braile, and C. L. Pruett. 2007. Vainly beating the air: Species-concept debates need not impede progress in science or conservation. Pages 30–44 *in* Festschrift for Ned K. Johnson: Geographic Variation and Evolution in Birds (C. Cicero and J. V. Remsen, Jr., Eds.). Ornithological Monographs, no. 63.

Zink, R. M., and N. K. Johnson. 1984. Evolutionary genetics of flycatchers. I. Sibling species in the genera *Empidonax* and *Contopus*. Systematic Zoology 33:205–216.

APPENDIX 1. Publications by Ned K. Johnson during his lifetime (121 papers and reviews, 2 monographs, 1 co-edited book, and 1 eight-authored 829-page book).

1949. JOHNSON, N. K. Loggerhead Shrike steals shot sparrow. Condor 51:233.
1950. JOHNSON, N. K. Slate-colored Junco in Reno, Nevada. Condor 52:134.
1952. JOHNSON, N. K. Additional records of the Rough-legged Hawk in Nevada. Condor 54:65.
1952. JOHNSON, N. K., AND F. RICHARDSON. Supplementary bird records for Nevada. Condor 54:358–359.
1953. JOHNSON, N. K. Dipper eaten by brook trout. Condor 55:158.
1954. JOHNSON, N. K. Notes on some Nevada birds. Great Basin Naturalist 14:15–18.
1954. JOHNSON, N. K. Food of the Long-eared Owl in southern Washoe County, Nevada. Condor 56:52.
1954. JOHNSON, N. K. New mammal records for Nevada. Journal of Mammalogy 35:577–578.
1956. JOHNSON, N. K. Recent bird record for Nevada. Condor 58:449–452.
1956. JOHNSON, N. K. Birds of the pinon association of the Kawich Mountains, Nevada. Great Basin Naturalist 16:32–33.
1958. JOHNSON, N. K. Notes on the Red Crossbill in Nevada. Condor 60:136–137.
1959. JOHNSON, N. K., AND R. C. BANKS. Pine Grosbeak and Lawrence Goldfinch in Nevada. Condor 61:303.
1961. JOHNSON, N. K., AND R. C. BANKS. A review of North American hybrid hummingbirds. Condor 63:3–28.
1962. JOHNSON, N. K., AND W. C. RUSSELL. Distributional data on certain owls in the western Great Basin. Condor 64:513–514.
1963. JOHNSON, N. K. Biosystematics of sibling species of flycatchers in the *Empidonax hammondii–oberholseri–wrightii* complex. University of California Publications in Zoology 66:79–238.
1963. JOHNSON, N. K., AND H. J. PEETERS. The systematic position of certain hawks in the genus *Buteo*. Auk 80: 417–446.
1963. JOHNSON, N. K. The supposed migratory status of the Flammulated Owl. Wilson Bulletin 75:174–178.
1963. JOHNSON, N. K. Comparative molt cycles in the tyrannid genus *Empidonax*. Pages 870–883 *in* Proceedings XIII International Ornithological Congress (C. G. Sibley, Ed.). American Ornithologists' Union, Washington, D.C.
1965. JOHNSON, N. K. The breeding avifaunas of the Sheep and Spring ranges in southern Nevada. Condor 67: 93–124.
1965. JOHNSON, N. K. Differential timing and routes of the spring migration in the Hammond Flycatcher. Condor 67:423–437.
1966. JOHNSON, N. K. Bill size and the question of competition in allopatric and sympatric populations of Dusky and Gray flycatchers. Systematic Zoology 15:70–87.
1966. JOHNSON, N. K. Morphologic stability versus adaptive variation in the Hammond Flycatcher. Auk 83: 179–200.
1969. JOHNSON, N. K. Review: Three papers on variation in flickers (*Colaptes*) by Lester J. Short, Jr. Wilson Bulletin 81:225–230.
1970. JOHNSON, N. K. The affinities of the boreal avifauna of the Warner Mountains, California. Occasional Papers of the Biological Society of Nevada 22:1–11.
1970. JOHNSON, N. K., AND C. M. EARHART. Size dimorphism and food habits of North American owls. Condor 72:251–264.
1970. JOHNSON, N. K. Fall migration and winter distribution of the Hammond Flycatcher. Bird-Banding 41: 169–190.
1970. JOHNSON, N. K., AND F. A. PITELKA. List of the Birds of the Berkeley Hills. Museum of Vertebrate Zoology, University of California, Berkeley.
1971. JOHNSON, N. K. Review: *Birds of the Lake Tahoe Region*, by R. T. Orr and J. Moffitt. Auk 90:224–225.
1971. JOHNSON, N. K. Ornithology at the Museum of Vertebrate Zoology, University of California, Berkeley. American Birds 25:537–538.
1972. JOHNSON, N. K. Origin and differentiation of the avifauna of the Channel Islands, California. Condor 74: 295–315.
1972. JOHNSON, N. K., AND A. H. BRUSH. Analysis of polymorphism in the Sooty-capped Bush Tanager. Systematic Zoology 21:245–262.
1972. JOHNSON, N. K. Breeding distribution and habitat preference of the Gray Vireo in Nevada. California Birds 3:73–78.
1973. JOHNSON, N. K. Spring migration of the Western Flycatcher, with notes on seasonal changes in sex and age ratios. Bird-Banding 44:205–220.
1973. JOHNSON, N. K. The distribution of boreal avifaunas in southeastern Nevada. Occasional Papers Biological Society of Nevada 36:1–4.

APPENDIX 1. Continued.

1973. AMERICAN ORNITHOLOGISTS' UNION. Thirty-second supplement to the American Ornithologists' Union Check-list of North American Birds. Auk 90:411–419.
1974. JOHNSON, N. K. Molt and age determination in Western and Yellowish flycatchers. Auk 91:111–131.
1974. JOHNSON, N. K., AND M. S. FOSTER. Notes on birds of Costa Rica. Wilson Bulletin 86:58–63.
1974. JOHNSON, N. K., AND K. L. GARRETT. Interior bird species expand breeding ranges into southern California. Western Birds 5:45–56.
1974. JOHNSON, N. K. Montane avifaunas of southern Nevada: Historical change in species composition. Condor 76:334–337.
1974. JOHNSON, N. K., AND J. F. LYNCH. Turnover and equilibria in insular avifaunas, with special reference to the California Channel Islands. Condor 76:370–384.
1975. JOHNSON, N. K. Controls of number of bird species on montane islands in the Great Basin. Evolution 29: 545–567.
1976. JOHNSON, N. K. Breeding distribution of Nashville and Virginia's warblers. Auk 93:219–230.
1976. JOHNSON, N. K., AND A. H. BRUSH. The evolution of color differences between Nashville and Virginia's warbler. Condor 78:412–414.
1976. AMERICAN ORNITHOLOGISTS' UNION. Thirty-third supplement to the American Ornithologists' Union Check-list of North American Birds. Auk 93:875–879.
1978. JOHNSON, N. K. Patterns of avian geography and speciation in the Intermountain Region. Great Basin Naturalist Memoirs 2:137–159.
1978. JOHNSON, N. K. Review: *Sexual Size Dimorphism in Hawks and Owls of North America*, by N. F. R. Snyder and J. W. Wiley. Wilson Bulletin 90:145–147.
1978. JOHNSON, N. K. Review: *Manual of Neotropical Birds, vol. 1*, by E. R. Blake. BioScience 28:460.
1980. JOHNSON, N. K. Character variation and evolution of sibling species in the *Empidonax difficilis–flavescens* complex (Aves: Tyrannidae). University of California Publications in Zoology, no. 112.
1980. JOHNSON, N. K. Vocal behavior of Euler's Flycatcher (*Empidonax euleri*) in Paraguay. Year Book of the American Philosophical Society 1979:212–213.
1982. AMERICAN ORNITHOLOGISTS' UNION. Thirty-fourth supplement to the American Ornithologists' Union Check-list of North American Birds. Supplement to the Auk 99:1CC–16CC.
1982. JOHNSON, N. K. Retain subspecies—At least for the time being. Auk 99:605–606.
1983. JOHNSON, N. K. Review: *The Avifauna of the South Farallon Islands, California*, by D. F. DeSante and D. G. Ainley. Auk 100:240–242.
1983. JOHNSON, N. K. Review: *Songs of the Vireos and Their Allies*, by Jon C. Barlow, narrated by J. W. Hardy. Ara Records. Wilson Bulletin 95:162–163.
1983. JOHNSON, N. K. Review: *The California Islands: Proceedings of a Multidisciplinary Symposium*, by D. M. Power (Ed.). Auk 100:782–784.
1983. AMERICAN ORNITHOLOGISTS' UNION. Check-list of North American Birds, 6th ed. American Ornithologists' Union, Washington, D.C.
1983. JOHNSON, N. K., AND R. M. ZINK. Speciation in sapsuckers (*Sphyrapicus*): I. Genetic differentiation. Auk 100:871–884.
1984. JOHNSON, N. K. Review: *The Hummingbirds of North America*, by P. A. Johnsgard. Auk 101:407–408.
1984. JOHNSON, N. K. List of the birds of the Berkeley Hills. Museum of Vertebrate Zoology, University of California, Berkeley.
1984. ZINK, R. M., AND N. K. JOHNSON. Evolutionary genetics of flycatchers. I. Sibling species in the genera *Empidonax* and *Contopus*. Systematic Zoology 33:205–216.
1984. JOHNSON, N. K., R. M. ZINK, G. F. BARROWCLOUGH, AND J. A. MARTEN. Suggested techniques for modern avian systematics. Wilson Bulletin 96:543–560.
1984. BARROWCLOUGH, G. F., N. K. JOHNSON, AND R. M. ZINK. On the nature of genic variation in birds. Pages 135–154 *in* Current Ornithology, vol. 2 (R. F. Johnston, Ed.). Plenum Press, New York.
1985. JOHNSON, N. K., AND C. B. JOHNSON. Speciation in sapsuckers (*Sphyrapicus*): II. Sympatry, hybridization, and mate preference in *S. ruber daggetti* and *S. nuchalis*. Auk 102:1–15.
1985. JOHNSON, N. K. Scientific natural history: Beyond cute little tricks. Review: Vertebrate Natural History, by M. F. Willson. BioScience 35:508–509.
1985. AMERICAN ORNITHOLOGISTS' UNION. Thirty-fifth supplement to the American Ornithologists' Union Check-list of North American Birds. Auk 102:680–686.
1985. JOHNSON, N. K., AND C. CICERO. The breeding avifauna of San Benito Mountain, California: Evidence for change over one-half century. Western Birds 16:1–23.

Appendix 1. Continued.

1985. Johnson, N. K., and R. M. Zink. Genetic evidence for relationships among the Red-eyed, Yellow-green, and Chivi vireos. Wilson Bulletin 97:421–435.

1985. DeSante, D. F., N. K. Johnson, R. LeValley, and R. P. Henderson. Occurrence and identification of the Yellow-bellied Flycatcher on Southeast Farallon Island, California. Western Birds 16:153–160.

1985. Johnson, N. K., and J. A. Marten. The geography of allelic frequencies in the Western Flycatcher (*Empidonax difficilis*). Pages 1029–1030 *in* Proceedings XVIII International Ornithological Congress (V. D. Ilyichev and V. M. Gavrilov, Eds.). Nauka Publishers, Moscow.

1986. Marten, J. A., and N. K. Johnson. Genetic relationships of North American cardueline finches. Condor 88:409–420.

1986. Johnson, N. K. Review: *Neotropical Ornithology*, by P. A. Buckley, M. S. Foster, E. S. Morton, R. S. Ridgeley, and F. G. Buckley (Eds.). Wilson Bulletin 98:616–618.

1986. Johnson, N. K., and C. Cicero. Richness and distribution of montane avifaunas in the White-Inyo Region, California. Pages 137–159 *in* Natural History of the White-Inyo Range, Eastern California and Western Nevada and High-altitude Physiology, vol. 1 (C. A. Hall, Jr., and D. J. Young, Eds.). University of California White Mountain Research Station, Bishop, California.

1987. American Ornithologists' Union. Thirty-sixth supplement to the American Ornithologists' Union Check-list of North American Birds. Auk 104:591–596.

1987. Johnson, N. K. Review: *The Naturalist's Field Journal*, by S. G. Herman. Auk 104:796–797.

1987. Zink, R. M., and N. K. Johnson. Avian biochemical systematics: A response to DeBenedictis. Birding 19: 17–20.

1988. Johnson, N. K., and J. A. Marten. Evolutionary genetics of flycatchers. II. Differentiation in the *Empidonax difficilis* complex. Auk 105:177–191.

1988. Johnson, N. K. Review: *Birds of the Great Basin: A Natural History*, by F. A. Ryser, Jr. Nevada Historical Society Quarterly 31:60–62.

1988. Johnson, N. K., R. M. Zink, and J. A. Marten. Genetic evidence for relationships in the avian family Vireonidae. Condor 90:428–445.

1988. Barrowclough, G. F., and N. K. Johnson. Genetic structure of North American birds. Pages 1630–1638 and 1669–1673 *in* Acta XIX Congressus Internationalis Ornithologici (H. Ouellet, Ed.). National Museum of Natural Sciences, University of Ottawa Press, Ottawa.

1989. Johnson, N. K. Review: *Speciation and Geographic Variation in Black-tailed Gnatcatchers*, by J. L. Atwood. Auk 106:348–349.

1989. Johnson, N. K., J. A. Marten, and C. J. Ralph. Genetic evidence for the origin and relationships of Hawaiian honeycreepers (Aves: Fringillidae). Condor 91:379–396.

1989. American Ornithologists' Union. Thirty-seventh supplement to the American Ornithologists' Union Check-list of North American Birds. Auk 106:532–538.

1989. Johnson, N. K. Review: *Avian Genetics: A Population and Ecological Approach*, by F. Cooke and P. A. Buckley (Eds.). Quarterly Review of Biology 64:345.

1989. Johnson, N. K. Review: *The Birds of Nevada*, by J. R. Alcorn. Auk 107:218–220.

1990. Johnson, N. K., and R. E. Jones. Geographic differentiation and distribution of the Peruvian Screech-Owl. Wilson Bulletin 102:199–212.

1991. Johnson, N. K., and J. A. Marten. Evolutionary genetics of flycatchers. III. Variation in *Empidonax hammondii* (Aves: Tyrannidae). Canadian Journal of Zoology 69:232–238.

1991. Johnson, N. K., and C. Cicero. Breeding birds. Pages 361–436 *in* California Natural History Guides, 55: Natural History of the White-Inyo Range, Eastern California (C. A. Hall, Jr., Ed.). University of California Press, Berkeley.

1991. American Ornithologists' Union. Thirty-eighth supplement to the American Ornithologists' Union Check-list of North American Birds. Auk 109:750–754.

1991. Johnson, N. K., and C. Cicero. Mitochondrial DNA sequence variability in two species of sparrows of the genus *Amphispiza*. Pages 600–610 *in* Acta XX Congressus Internationalis Ornithologici (B. D. Bell, Ed.). Congress Trust Board, Wellington, New Zealand.

1992. Johnson, N. K. Review: *Utah Birds: Historical Perspectives and Bibliography*, by W. H. Behle. Auk 109: 210–211.

1992. Johnson, N. K., and J. A. Marten. Macrogeographic patterns of morphometric and genetic variation in the Sage Sparrow complex. Condor 94:1–19.

1992. Cicero, C., and N. K. Johnson. Genetic differentiation between populations of Hutton's Vireo (Aves: Vireonidae) in disjunct allopatry. Southwestern Naturalist 37:344–348.

APPENDIX 1. Continued.


1993. JOHNSON, N. K., AND R. E. JONES. The Green Jay turns blue in Peru: Interrelated aspects of the annual cycle in the arid tropical zone. Wilson Bulletin 105:388–398.

1993. JOHNSON, N. K. Review: *Molt of the Genus* Spizella *(Passeriformes, Emberizidae) in Relation to Ecological Factors Affecting Plumage Wear*, by E. J. Willoughby. Wilson Bulletin 105:541–542.

1993. AMERICAN ORNITHOLOGISTS' UNION. Thirty-ninth supplement to the American Ornithologists' Union Check-list of North American Birds. Auk 110:675–682.

1993. JOHNSON, N. K. Review: *Handbook of the Birds of the World*, vol. 1, by J. del Hoyo, A. Elliott, and J. Sargatal (Eds.). Auk 110:959–960.

1994. JEHL, J. R., AND N. K. JOHNSON, EDS. A century of avifaunal change in western North America. Studies in Avian Biology, no. 15.

1994. JOHNSON, N. K., AND J. R. JEHL, JR. A century of avifaunal change in western North America: Overview. Pages 1–3 *in* A Century of Avifaunal Change in Western North America (J. R. Jehl, Jr., and N. K. Johnson, Eds.). Studies in Avian Biology, no. 15.

1994. JOHNSON, N. K. Pioneering and natural expansion of breeding distributions in western North American birds. Pages 27–44 *in* A Century of Avifaunal Change in Western North America (J. R. Jehl, Jr., and N. K. Johnson, Eds.). Studies in Avian Biology, no. 15.

1994. ANONYMOUS. Elliott Coues Award, 1993: Joel L. Cracraft. Auk 111:239–240.

1994. JEHL, J. R., JR., AND N. K. JOHNSON. 1994. Review: *The Birds of North America: Life Histories for the 21st Century, vol. 1*, by A. Poole, P. Stettenheim, and F. Gill (Eds.). Condor 96:833–839.

1994. JOHNSON, N. K. Old-school taxonomy versus modern biosystematics: Species-level decisions in *Stelgidopteryx* and *Empidonax*. Auk 111:773–780.

1994. CICERO, C., AND N. K. JOHNSON. Mitochondrial DNA divergence at the cytochrome *b* locus in sapsuckers. Journal für Ornithologie 135:351.

1995. JOHNSON, N. K. Seven avifaunal censuses spanning one-half century on an island of white firs (*Abies concolor*) in the Mojave Desert. Southwestern Naturalist 40:76–85.

1995. JOHNSON, N. K. Ornithology at the Museum of Vertebrate Zoology, University of California, Berkeley. Pages 183–221 *in* Contributions to the History of North American Ornithology (W. E. Davis, Jr., and J. A. Jackson, Eds.). Memoirs of the Nuttall Ornithological Club, no. 12.

1995. JOHNSON, N. K. Speciation in vireos. I. Macrogeographic patterns of allozymic variation in the *Vireo solitarius* complex in the contiguous United States. Condor 97:903–919.

1995. AMERICAN ORNITHOLOGISTS' UNION. Fortieth supplement to the American Ornithologists' Union Check-list of North American Birds. Auk 112:819–830.

1995. CICERO, C., AND N. K. JOHNSON. Speciation in sapsuckers (*Sphyrapicus*): III. Mitochondrial-DNA sequence divergence at the cytochrome-*b* locus. Auk 112:547–563.

1996. JOHNSON, N. K. Review: *Handbook of the Birds of the World, vol.* 2, by J. del Hoyo, A. Elliott, and J. Sargatal (Eds.). Auk 113:518–519.

1996. CICERO, C., AND N. K. JOHNSON. Declining scientific standards in studies of avian distribution. [Review of *California Birds: Their Status and Distribution*, by A. Small.] Auk 113:522–523.

1997. AMERICAN ORNITHOLOGISTS' UNION. Forty-first supplement to the American Ornithologists' Union Check-list of North American Birds. Auk 114:542–552.

1998. JOHNSON, N. K., J. V. REMSEN, JR., AND C. CICERO. Refined colorimetry validates endangered subspecies of the Least Tern. Condor 100:18–26.

1998. BANKS, R. C., J. W. FITZPATRICK, T. R. HOWELL, N. K. JOHNSON, B. L. MONROE, JR., H. OUELLET, J. V. REMSEN, AND R. W. STORER. Check-list of North American Birds, 7th ed. Allen Press, Lawrence, Kansas.

1998. CICERO, C. AND N. K. JOHNSON. Molecular phylogeny and ecological diversification in a clade of New World songbirds (genus *Vireo*). Molecular Ecology 7:1359-1370.

1998. JOHNSON, N. K., J. V. REMSEN, JR., AND C. CICERO. Biologic species of birds in theory and practice. Abstract *in* Proceedings of the XXII International Ornithological Congress, Durban (N. J. Adams and R. H. Slotow, Eds.). Ostrich 69:81.

1999. JOHNSON, N. K., J. V. REMSEN, JR., AND C. CICERO. Resolution of the debate over species concepts in ornithology: A new comprehensive biologic species concept. Pages 1470–1482 *in* Acta XXII Congressus Internationalis Ornithologici (N. J. Adams and R. H. Slotow, Eds.). BirdLife South Africa, Johannesburg.

2000. JOHNSON, N. K. Review: *The Era of Allan R. Phillips: A Festschrift*, compiled by R. W. Dickerman. Wilson Bulletin 112:157–158.

2001. JOHNSON, N. K., AND R. E. JONES. A new species of tody-tyrant (Tyrannidae: *Poecilotriccus*) from northern Peru. Auk 118:334–341.

APPENDIX 1. Continued.

2001. JOHNSON, N. K. Review: *The Directory of Australian Birds: Passerines,* by R. Schodde and I. J. Mason. Condor 103:200.

2001. MAYR, E., AND N. K. JOHNSON. Commentary: Is *Spizella taverneri* a species or a subspecies? Condor 103: 418–419.

2001. CICERO, C., AND N. K. JOHNSON. Higher level phylogeny of New World vireos (Aves: Vireonidae) based on sequences of multiple mtDNA genes. Molecular Phylogenetics and Evolution 20:27–40.

2001. HOPENSTAND, D., AND N. K. JOHNSON. Variation in pigmentary and colorimetric properties of red crown feathers in selected woodpeckers (Aves: Picidae). Berkeley Scientific 5:56–60.

2001. JOHNSON, N. K. The role of pigments and structural colors in the evolutionary origin of feathers. Evolutionary Theory 12:91–97.

2002. CICERO, C., AND N. K. JOHNSON. Phylogeny and character evolution in the *Empidonax* group of tyrant flycatchers (Aves: Tyrannidae): A test of W. E. Lanyon's hypothesis using mtDNA sequences. Molecular Phylogenetics and Evolution 22:289–302.

2002. JOHNSON, N. K. Leapfrogging revisited in Andean birds: Geographic variation in the tody-tyrant superspecies *Poecilotriccus ruficeps* and *P. luluae*. Ibis 144:69–84.

2002. JOHNSON, N. K., AND C. CICERO. The role of ecologic diversification in sibling speciation of *Empidonax* flycatchers (Tyrannidae): Multigene evidence from mtDNA. Molecular Ecology 11:2065–2081.

APPENDIX 2. Master's (3) and Ph.D (15) students sponsored or cosponsored (*) by Ned K. Johnson, excluding students sponsored at the time of his death in 2003.

| | | |
|---|---|---|
| 1965 | George T. Ferrell, MA | Variation in blood group frequencies in populations of Song Sparrows of the San Francisco Bay Region |
| 1965 | Mercedes Foster, MA | Pterylography, molt, and age determination of the Orange-crowned Warbler. |
| 1969 | Robert B. Hamilton, Ph.D | Comparative behavior of the American Avocet (*Recurvirostra americana*) and the Black-winged Stilt (*Himantopus h. mexicanus*) |
| 1972 | Luis F. Baptista, Ph.D | Demes, disperson and song dialects of sedentary populations of the White-crowned Sparrow (*Zonotrichia leucophyrys nuttalli*) |
| 1973 | Richard E. Johnson, Ph.D* | Biosystematics of the avian genus *Leucosticte* |
| 1974 | James F. Lynch, Ph.D* | Ontogenetic and geographic variation in the morphology and ecology of the black salamander (*Aneides flavipunctatus*) |
| 1977 | Carolyn S. Connors, MA | Foraging ecology of Black Turnstones and Surfbirds on their wintering grounds at Bodega Bay, California |
| 1978 | Stephen F. Bailey, Ph.D | Foraging strategies of frugivorous birds in relation to the availability of berries, with special reference to central California |
| 1983 | Robert M. Zink, Ph.D | Patterns and evolutionary significance of geographic variation in the schistacea group of the Fox Sparrow (*Passerella iliaca*) (Oregon, Nevada, California) |
| 1990 | Jeffrey G. Groth, Ph.D | Cryptic species of nomadic birds in the Red Crossbill (*Loxia curvirostra*) complex of North America |
| 1992 | Douglas A. Bell, Ph.D | Hybridization and sympatry in the Western Gull/Glaucous-winged Gull complex |
| 1992 | Scott V. Edwards, Ph.D* | Gene flow and mitochondrial DNA evolution in social babblers (Aves: *Pomatostomus*) |
| 1993 | Carla Cicero, Ph.D* | Sibling species of titmice in the *Parus inornatus* complex (Aves: Paridae) |
| 1993 | Julia I. Smith, Ph.D | Environmental influence on the ontogeny, allometry, and behavior of the Song Sparrow (*Melospiza melodia*) |
| 1996 | Kevin J. Burns, Ph.D | Molecular phylogenetics of tanagers and the evolution of sexual dimorphism in plumage |

Appendix 2. Continued.

| | | |
|---|---|---|
| 2001 | Thomas A. Stidham, Ph.D* | The origin and ecological diversification of modern birds: Evidence from the extinct wading ducks, Presbyornithidae (Neornithes: Anseriformes) |
| 2002 | Alison L. Chubb, Ph.D | Molecular evolution and phylogenetic utility of the ZENK gene in birds |
| 2002 | Jason A. Mobley, Ph.D | Molecular phylogenetics and the evolution of nest building in kingbirds and their allies (Aves: Tyrannidae) |

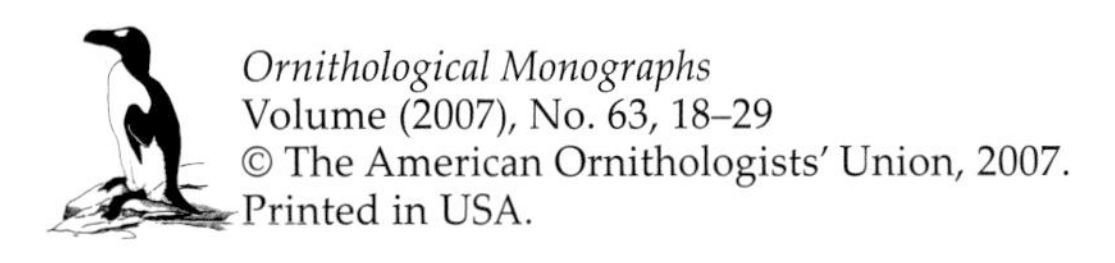
*Ornithological Monographs*
Volume (2007), No. 63, 18–29


CHAPTER 2

# MOLECULAR ADVANCES IN THE STUDY OF GEOGRAPHIC VARIATION AND SPECIATION IN BIRDS

ALLAN J. BAKER[1]

*Department of Ecology and Evolutionary Biology, University of Toronto, Toronto, Ontario M5S 3B2, Canada*

ABSTRACT.—Problems in deciphering the patterns and causes of geographic variation and speciation in birds occupied Ned Johnson (e.g., Johnson 1980, Cicero and Johnson 1998, Johnson and Cicero 2002) and many other ornithologists for much of their lives, but the recent onslaught of molecular studies and associated analytical methods are providing breakthroughs in understanding these evolutionary phenomena. In particular, coalescent theory and Markov chain Monte Carlo (MCMC) applications have shown that bird species are sometimes strongly structured into well-differentiated populations by historical subdivision, high philopatry, and small effective population sizes, whereas other species that have recently recolonized parts of their range are effectively panmictic. These are the sorts of results that were impossible to obtain from studies of geographic variation in phenotypic characters alone. Recovery of well-supported species trees from gene trees is much more likely when multiple genes are sequenced, and provides the means for inferring divergence times and patterns and processes of evolution in birds. As in other vertebrates, patterns of cladogenesis in large clades of birds correlate with major paleoenvironmental changes and associated adaptive radiations, reminding us that much of current biodiversity on the planet had its genesis in the distant past.
*Received 24 July 2006, accepted 5 February 2007.*

RESUMEN.—Ornitólogos como Ned Johnson, entre muchos otros, han dedicado gran parte de sus carreras a analizar las causas y patrones en la variación geográfica y la especiación en aves (e.g. Johnson 1980, Cicero and Johnson 1998, Johnson and Cicero 2002). Recientemente, los estudios moleculares y los métodos analíticos asociados han provisto de las herramientas necesarias para entender estos fenómenos evolutivos. En particular, el uso de la teoría de coalescencia y los algoritmos asociados a las cadenas de Markov Monte Carlo han demostrado que algunas especies de aves presentan una estructura poblacional muy marcada, producto de subdiviciones históricas, filopatría y tamaños poblacionales efectivos reducidos, mientras que otras especies que recientemente han recolonizado una parte de su área de distribución son panmíticas. Estos resultados son difíciles de obtener estudiando únicamente la variación geográfica de caracteres fenotípicos. Al igual que en otros vertebrados, los patrones de cladogénesis en grandes clados de aves se correlacionan con eventos paleoambientales y radiaciones adaptativas asociadas, recordándonos que gran parte de la biodiversidad del planeta tuvo su génesis en un pasado lejano.

NED JOHNSON HAD a consuming interest in the study of geographic variation and its connection to speciation in birds and was a strong advocate for the biological species concept (e.g., Johnson et al. 1999). He and his students and postdocs embraced new methods of detecting and quantifying variation within and among species as the way to make groundbreaking progress in this ever-evolving field and, thus, his contributions were many and varied. A landmark paper from Johnson's laboratory on the nature of genic variation in birds established the importance of neutrality of molecular markers in analyzing genetic structure in avian species (Barrowclough et al. 1985). Subsequent use of putatively neutral markers elucidated some of the major forces acting on avian populations, including founder effects and bottlenecks, genetic drift, gene flow, and geographic

[1]Present address: Department of Natural History, Royal Ontario Museum, Toronto, Ontario M5S 2C6, Canada. E-mail: allanb@rom.on.ca

isolation (e.g., Baker and Moeed 1987, Wenink et al. 1996, Clegg et al. 2002). Although ornithologists have used molecular methods to elucidate the timing, geography and ecology of speciation, the molecular and behavioral mechanisms of speciation await thorough investigation in birds (for an excellent review, see Edwards et al. 2005). The roles of genetic incompatibilities and behavioral mechanisms in speciation, including sexual selection, song learning, imprinting, and reinforcement, are currently hot topics (e.g., Irwin et al. 2001, Sorenson et al. 2003).

As a tribute to the achievements of Johnson and his collaborators and the impact they have had on the field, it is appropriate to highlight some of the advances being made in the study of population structure and modes of speciation in birds, especially with respect to molecular tools and their applications. I have focused on a very selective set of topics, with exemplars mainly from my own laboratory. To make some specific points, I outline a few examples of molecular population structure in bird species, discuss the roles of isolation and philopatry in geographic variation and speciation, and point out the role of ancient paleoenvironmental changes in generating species radiations that are reflected in extant avian biodiversity.

## Population Genetic Structure in Birds

### A Need for Multiple Markers

Although it has long been believed that populations of most bird species may be panmictic or only weakly structured (e.g., Zink 1997) because of homogenizing gene flow mediated by the dispersal power of flight, molecular assays have revealed increasingly that many widely distributed populations are subdivided (e.g., Wenink et al. 1996, Irwin et al. 2001, Griswold and Baker 2002, Zink 2004, Baker et al. 2005, Brito 2005). Use of DNA-sequence data sets has not only provided genetic markers for detecting population structure in avian species, but has also catalyzed analytical advances via mathematical models (e.g., coalescent theory) that give exact probabilities of the data as a function of the parameters that underlie population history (Hein et al. 2004).

This approach is illustrated below for an ancestral population that split into two populations, some time in the past, which potentially exchange migrants (Nielsen and Wakeley 2001). Control-region (CR) sequences of 403 base pairs (bp) (Baker et al. unpubl. data) obtained from samples of two putative subspecies of the Purple Martin from east (*Progne subis subis*; $n$ = 71) and west (*P. s. arboricola*; $n$ = 50) of the Rocky Mountains were analyzed using the program MDIV (Nielsen and Wakeley 2001), which implements this basic non-equilibrium model. Modal values of the population mutation parameter ($\theta$ = 5.95 substitutions locus$^{-1}$), time of population divergence ($tpop$ = 2.72), time to the most recent common ancestor (TMRCA) and scaled migration rate ($M$ = 0.13) were obtained from the posterior distributions generated by the MCMC procedure after 2 million generations and a burn-in period of 500,000 generations (Fig. 1). To convert these values to time in years, we estimated generation time using the equation (Sæther et al. 2005) $g = \alpha \times (s/1 - s)$, where $\alpha$ is age at first breeding and $s$ is annual

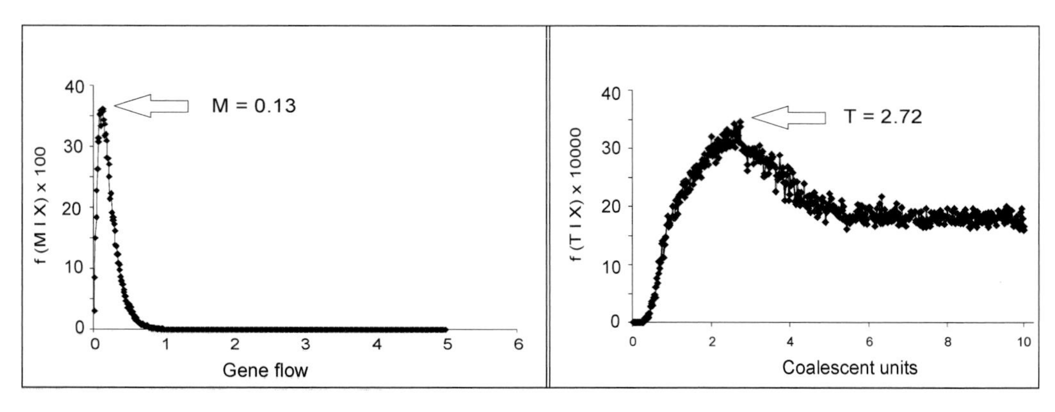


Fig. 1. Posterior distributions of scaled migration rate ($M$) and population divergence time ($Tpop$) generated in MDIV for eastern and western subspecies of Purple Martins.

survival (0.7). This yielded a generation time of 3.3 years for Purple Martins. We also estimated mutation rate ($\mu$) for the CR sequences using a range of calculated mutation rates for the CR in birds (5%, 10%, and 20% per million years; Brito 2005) because a specific rate for Purple Martins is not available. For example, a rate of 10% per million years converts to $10^{-7}$ substitutions per site per year and a per-gene rate of $3.3 \times 403 \times 10^{-7}$ substitutions per generation. Population divergence time can be calculated with the expression $([tpop \times \theta]/2) \times 1/\mu = 60{,}800$ generations, or ~200,400 years before present (ybp). Similarly, mutation rates of 5% per million years and 20% per million years translate to a range of about 400,000 to 100,000 ybp for the population splitting time. Using a mutation rate of 10% per million years, for example, the TMRCA of the CR sequences in the gene tree is 3.16, which translates to ~230,000 ybp. Because the TMRCA can be used to approximate the age of the ancestral population, we can infer that migration between the two descendant subspecies populations probably ceased shortly after their historical separation.

However, mitochondrial DNA (mtDNA) is only one gene and, thus, the 95% confidence intervals on demographic parameters are substantial. Recently, major concerns have been raised about inferences of historical demography based solely on this genome, owing to its unusual evolutionary rules and the influence of selection (Ballard and Whitlock 2004, Bazin et al. 2006). For example, the discovery of low mtDNA diversity and higher nuclear-DNA diversity in warblers of the genus *Phylloscopus* suggests that maternally inherited genes were affected by a selective sweep (Bensch et al. 2006).

One solution to these potential problems is to assay molecular variation at multiple unlinked nuclear loci to provide independent estimates of the within-species genealogy. This enables checking of inferences made from mtDNA sequences alone and provides reduced confidence intervals around demographic parameters. Even a comparison of phylogeographic patterns based on mtDNA sequences versus microsatellites often can be instructive, as for example in the Brown Kiwi (*Apteryx australis*) in New Zealand (Burbidge et al. 2003; see Fig. 2). At least in this case, the genealogies appear to be very similar, and thus the CR sequences and the nine microsatellite loci could be combined in a coalescent program such as IM (Hey and Nielsen 2004) to improve the inference of demographic parameters for South Island versus Stewart Island. A recent innovative example in Jennings and Edwards (2005) showed how a multilocus coalescent approach can make advances in understanding the evolutionary connection between geographic variation and speciation. They analyzed variation in 30 anonymous nuclear loci of three closely related species of Australian grass-finches (*Poephila* spp.) and, despite different topologies and coalescent times for each gene tree, they had enough information to estimate population divergence times and ancestral population sizes.

### Determining the Number of Populations

The advent of molecular analyses of samples collected over large stretches of geography have forced investigators to confront again the problem faced earlier by ornithologists studying geographic variation in classical phenotypic traits: how do we determine the number of populations represented by a number of samples? Often, the conventional solution to this problem has been to use the sampling locale as the "population" for analysis, but newer methods, such as those in the programs BAPS (Corander et al. 2003) and STRUCTURE (Pritchard et al. 2000), can infer population structure using multilocus data. For example, Given (2004) assayed variation in 10 microsatellite loci from samples of Silver Gulls (*Larus novaehollandiae*) collected at nine locales in mainland New Zealand ($n = 224$) and one locale on sub-Antarctic Campbell Island ($n = 24$). Analysis with both clustering methods placed all New Zealand samples in one population and Campbell Island in another population (Fig. 3). Nevertheless, distance between sampled colonies and their genetic differentiation (*Fst*) were positively correlated (Mantel test; $Z = -209.18$, $r^2 = 0.46$, $P = 0.004$; Given 2004). This indicates that gene flow between colonies fits an isolation-by-distance model and is sufficient to prevent significant population structure in the New Zealand mainland but not between New Zealand and geographically remote Campbell Island.

Caution is warranted, however, in using multilocus data to interpret the biological meaning of statistically significant population structure in birds, as illustrated in the analysis of 157

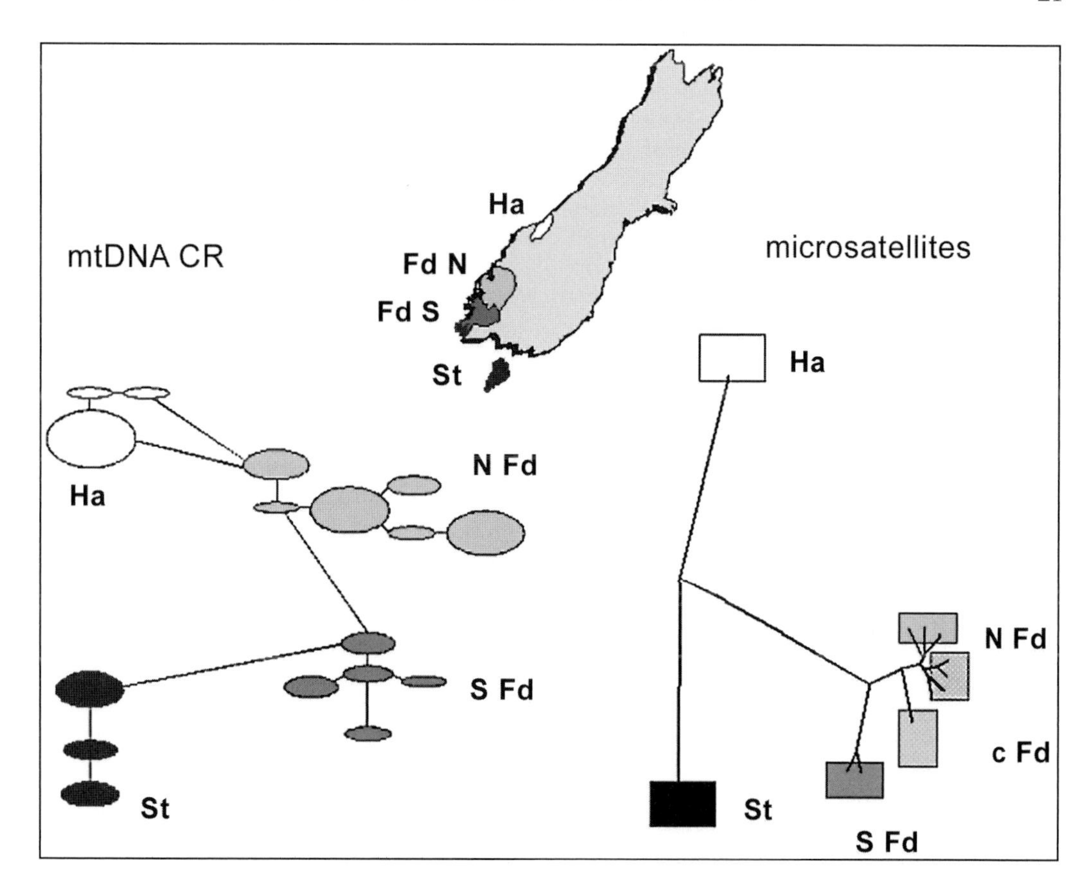


FIG. 2. Genetic divergence in mtDNA CR sequences and nine microsatellite loci among four regional populations of *Apteryx australis* in the South Island of New Zealand (Burbidge et al. unpubl. data).

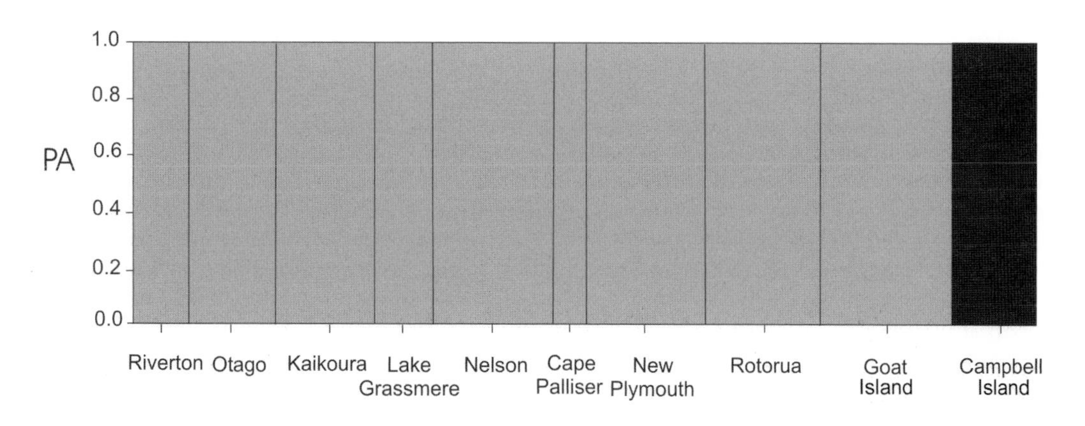


FIG. 3. Identification of populations of *Larus novaehollandiae* in the New Zealand region based on the estimated proportion of ancestry (PA) of individuals using 10 microsatellite loci with the program STRUCTURE. From Given (2004).

polymorphic nuclear loci identified by amplified fragment-length polymorphisms in populations of House Finches (*Carpodacus mexicanus frontalis*) in North America (Wang et al. 2003). Analysis of this data set with STRUCTURE diagnosed three almost completely non-overlapping populations that might be assigned naively to three species under a version of the phylogenetic species concept, one in their native range in the western United States and two others introduced separately to Hawaii and the eastern United States within the past 100 years. Instead, this example indicates that significant shifts in allele frequencies can occur rapidly through genetic drift and selection.

### Geographic Variation and Speciation

When pronounced patterns of geographic variation are detected in samples, it is often difficult to decide whether they represent population structure within a species or whether they are sufficiently discrete to be considered different species. This is especially true in analyzing patterns of morphological variation and is further accentuated in extinct organisms, for which it is impossible to judge the degree of genetic incompatibility among well-differentiated samples. A telling example is provided by the extinct moas of New Zealand, which are known from a large number of subfossil remains recovered from swamps, caves, and middens. The complex nature of morphological variation led to the naming of at least 64 species and 20 genera of moa by early workers, which was whittled down to 11 species in 6 genera in the 1990 *Checklist of New Zealand Birds* (Turbott 1990). However, when it proved possible to amplify nuclear genes for molecular sexing from ancient DNA extracted from within the cortex of fossil bones, an unexpectedly large degree of sexual dimorphism was discovered in most genera, with smaller bones belonging to males and larger ones to females (with some overlap at intermediate sizes; Bunce et al. 2003, Huynen et al. 2003). Males and females from one species of the giant *Dinornis* had been split into two species using bone size alone, because they differed in size by up to 50%.

To test the prevailing taxonomy and to attempt to partition morphological variation into geographic variation within versus among species, 125 specimens in museum collections were typed genetically with a 658 bp mtDNA CR sequence (Baker et al. 2005). Bayesian analysis of these sequences recovered 14 monophyletic lineages, 9 of which are currently recognized, plus 5 new lineages that may warrant species status. When exemplars of these 14 lineages were sequenced for a total of 2,814 bp from 10 mtDNA genes (including the CR sequences), a strongly supported phylogeny was derived. Molecular dating with a relaxed molecular clock to account for evolutionary-rate variation among lineages indicated that the most recent common ancestor of moas can be traced back to just after the Oligocene "drowning" of New Zealand, when previous populations likely underwent severe reductions. Phylogeographic patterns in gene trees can be highly informative with respect to the mode of speciation, because they usually bear the signature of the geographic and population-demographic or selective events that accompany speciation (Avise 2000). Much of the cladogenesis in the CR gene tree is concentrated in the period around 4–10 mya, which corresponds to the time when the landmass was fragmented by tectonic and mountain-building events as well as a globally cooling climate (Fig. 4). Lineages therefore became isolated geographically and probably speciated allopatrically with concomitant ecological specialization.

The discovery of the moa lineages using ancient DNA provided an opportunity to test the efficacy of DNA barcoding with the standard cytochrome oxidase I (COI) sequence. This technique has been criticized because it needs to be evaluated in a phylogenetic framework in the limiting case of sister-species comparisons (Moritz and Cicero 2004) or it fails on different clades of organisms (e.g., Meier et al. 2006) or in broad parameter space (Hickerson et al. 2006). Twelve of the 14 lineages that were known at the time of testing had DNA barcodes that differed by >2.7% (Lambert et al. 2005), which exceeds the 10× threshold for within-species differences calibrated for North American birds (Hebert et al. 2004). Subsequently, the two new lineages of *Dinornis* discovered in the South Island of New Zealand were found to have nearly identical COI sequences but were easily distinguished by their CR sequences and had been accorded separate taxonomic status in the past on the basis of skeletal differences. Given the recency of their divergence times, the slow rate of mtDNA evolution in moas, and the correlation done of

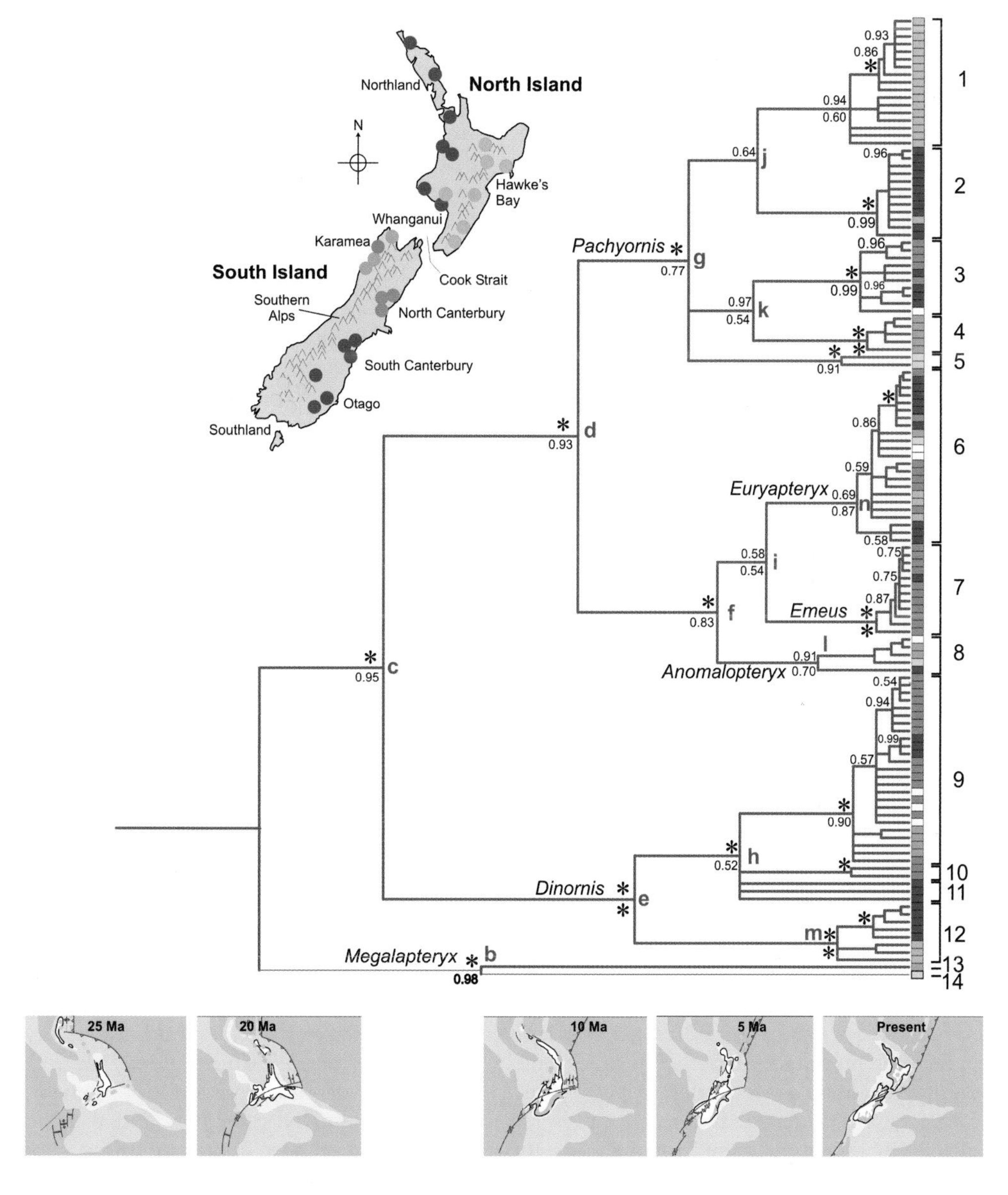


FIG. 4. Bayesian tree constructed with 658-bp CR sequences from 125 moa specimens under a GTR+I+G model of evolution. Numbers at the branch tips identify the 14 major lineages as follows: (1) *Pachyornis mappini*, (2) *P. n.sp.A*, (3) *P. elephantopus*, (4) *P. australis*, (5) *P. n.sp.B.*, (6) *Euryapteryx geranodies*, (7) *Emeus crassus*, (8) *Anomalopteryx didiformis*, (9) *Dinornis robustus*, (10) *D. n.sp.A*, (11) *D. n.sp.B*, (12) *D. novaezealandiae*, (13) *Megalapteryx didinus*, and (14) *M. benhami*. Specimens are color-coded according to geographic locations plotted along with place names on the map. From Baker et al. (2005). Samples of lineages (1) and (2) previously referred to *P. mappini* were used for coalescent modeling with the program IM.

percentage of divergence of COI barcodes with time of lineage-splitting (Tavares and Baker unpublished data), it may be more appropriate to barcode this clade of birds with the central conserved domain of the CR.

The geographic partitioning of the two lineages of *Pachyornis mappini* in the eastern ($n$ = 17) and western ($n$ = 12) regions of the North Island of New Zealand (Fig. 4) suggests that they were effectively isolated a long time ago. To estimate their time of splitting, a coalescent analysis was run in IM (Hey and Nielsen 2004), which relaxes the assumption of equal population size in the ancestral and two daughter populations and of symmetric gene flow between populations inherent in MDIV. Two runs of IM were conducted, each for 7 million generations with priors for maximum effective population size ($N_1$ and $N_2$) and "population" splitting time ($t$) set to 30, and maximum migration rates ($m_1$ and $m_2$) between the two populations set to 10. Convergence of the four chains in the MCMC was monitored by autocorrelations (<0.03), and effective sample-size estimates (>235) were generated by the program for all the parameters. Similar estimates were obtained in both runs, with scaled population-size estimates for the two populations ($q_1 = 56.95$, $q_2 = 24.45$), and ancestral population size $q_A = 26.99$), coalescent time ($t = 12.00$), and migration rates into population 1 ($m_1 = 0.005$) and population 2 ($m_2 = 0.005$). To convert these parameter values to effective population sizes of females, time in years, and gene flow per generation, two estimates of mutation rate were used for the conserved central domain of the CR of these moas. Rates of molecular evolution in ratites are generally slower than in other birds (Pereira and Baker 2006), so it may be appropriate to use rates of 2.5% and 5% per million years as observed in some slower-evolving Neoaves (Table 1).

The low levels of migration and the potentially deep time of population splitting between these lineages indicate that they evolved in isolation, on opposite sides of the North Island separated by mountain barriers. Effective population sizes of females are moderate, and assuming a 1:1 sex ratio and that census population size was ~5× larger (as in the Brown Kiwis), the two populations of moa could have comprised 200,000–400,000 individuals. These rough estimates support the claim by Gemmell et al. (2004) that moa populations in New Zealand were large. However, demographic parameters estimated with the single CR locus also need to be estimated with multiple nuclear genes to improve confidence and to set narrower 90% bounds on the posterior distributions. For example, estimation of 90% highest posterior density interval (90% HPD) values for ancestral effective population size was not possible, but given the wide 90% HPD values for $N_1$ and $N_2$ that overlap the modal estimate for $N_A$, this suggests that modeling the moa sequences under constant population size (as was done in this example) is reasonable. Coalescent modeling of the sequences in DNAsp (Rozas et al. 2003) also failed to detect population growth ($Fs = -2.52$, $P = 0.19$).

### Influence of Paleoenvironmental Changes on Speciation in Birds

The blossoming of studies of variation in DNA sequences has not only clarified the nature of population structure in bird species but also has opened new vistas in avian evolution. In particular, phylogenetic studies of major clades of birds have shown that patterns of cladogenesis and speciation have their genesis in deep time (e.g., Haddrath and Baker 2001, Paton et al. 2002, Pereira and Baker 2006, Tavares et al. 2006) and, thus, may be connected causally to major paleoenvironmental changes in the past. This perspective has emerged because investigators have increasingly employed large DNA-sequence data sets from multiple genes from nuclear and mtDNA genomes, and also because advances have been made in fitting separate

Table 1. Estimates of demographic quantities and their 90% highest posterior density intervals (90% HPD) for two lineages of *P. mappini* using a per-locus mutation rate of 5% per million years. Halving this rate will double estimates of effective population sizes and population divergence time.

| Demographic parameter | Mode | Lower 90% HPD | Upper 90% HPD |
|---|---|---|---|
| $N_1$ | 43,278 | 23,518 | 83,870 |
| $N_2$ | 18,578 | 7,625 | 42,204 |
| $N_A$ | 20,511 | – | – |
| $t$ (years) | 364,589 | 131,762 | 902,279 |
| $2N_1m_1 = \theta_1m_1/2$ | 0.14 | 0.14 | 3.27 |
| $2N_2m_2 = \theta_2m_2/2$ | 0.06 | 0.06 | 2.88 |

models of sequence evolution to each gene partition or codon position. Bayesian methods (e.g., Ronquist and Hulsenbeck 2003) have speeded this transition and have facilitated the use of relaxed molecular clocks (e.g., Thorne and Kishino 2002) that accommodate variation in evolutionary rate among species and account for other major sources of phylogenetic uncertainty. This has allowed estimation of clade and species divergence times and their 95% confidence intervals, thereby enabling comparisons with the onset of well-documented paleoenvironmental changes.

Apart from the moa example outlined above, penguins and Neotropical parrots provide two additional examples of the influence of large-scale paleoenvironmental changes. A strongly supported phylogeny constructed from 2,802 bp of the nuclear gene RAG-1 and 2,899 bp of the mitochondrial genes 12S, 16S, cytochrome *b*, and COI was used to derive a Bayesian estimate of the ancestral area for modern penguins in Antarctica (Baker et al. 2006). To explain the current circumpolar distribution and restriction of penguins to the southern oceans, a chronogram was produced that depicts estimates of divergence times at the nodes and associated 95% credibility intervals (Fig. 5). Major global cooling events and the formation of ice sheets in Antarctica were shown to coincide with two bouts of cladogenesis in penguins. The first cooling period (about 34–25 mya) coincided with the divergence of three genera (*Spheniscus, Eudyptes,* and *Eudyptula*) with temperate-latitude distributions from older Antarctic genera. Their dispersal northwards out of Antarctica was probably via the newly formed circumpolar current. A second bout of diversification leading to the rise of multiple species of extant penguins dated from 12–14 mya at the time of the middle Miocene climate transition (MMCT), when surface waters in the southern oceans cooled further and ice volume in Antarctica increased again. Because the younger species resulting from this climate-induced expansion out of Antarctica are distributed today on isolated islands and the tips of southern continents, we can infer that they probably speciated allopatrically. Only the species in the genera *Aptenodytes* and *Pygoscelis* stayed in Antarctica and adapted to cooler conditions, but they too may have left the continent during glacial maxima (Baker et al. 2006).

A second example is nicely demonstrated by Neotropical parrot genera (Tavares et al. 2006). Using 6,388 bp from RAG-1 and seven mtDNA genes, they constructed a strongly supported phylogeny of 29 species in 25 of the 30 genera. Amazons and their allies were resolved as a sister clade to macaws, conures, and allies, which are jointly sister to parrotlets (Fig. 6). Molecular dating with relaxed clocks placed the divergence of Neotropical and Australian parrots between 51 and 66 mya, thus indicating that ancestral parrots were widespread in Gondwanaland before fragmentation by continental drift. The three major clades of Neotropical parrots were estimated to have diverged between 41 and 57 mya, but patterns of speciation are quite different within these clades.

Genera of amazons and allies diversified steadily between 46 and 16 mya, probably in South America, as forests expanded and contracted in response to global shifts in temperature and sea levels that periodically fragmented the biota. By contrast, macaws, conures, and their relatives radiated much later, beginning ~28 mya, when ecological speciation would have been mediated by the uplift of the Andes and by the formation of dry grassland habitats that they invaded from the forests. The two parrotlet genera may be ancient relicts that originated from an ancestral stock in Antarctica or the southern cone of South America and were driven northward by the formation of ice sheets in Antarctica.

## Conclusions

The age of population genomics is just dawning, and we can look forward to many more developments in the near future as molecular technologies continue to advance. Already, new sequencing technologies have demonstrated how even highly fragmented ancient DNA in Neanderthals and mammoths can be retrieved and possibly used to reconstruct whole genomes (Green et al. 2006, Poinar et al. 2006). Genome-sequencing costs are likely to decrease significantly in the near future. This will enable targeted approaches on a multitude of genes, leading not only to much more reliable estimates of population divergence times, gene flow, and effective population size, but also a new age in phylogenomics and genetic mechanisms of speciation in birds.

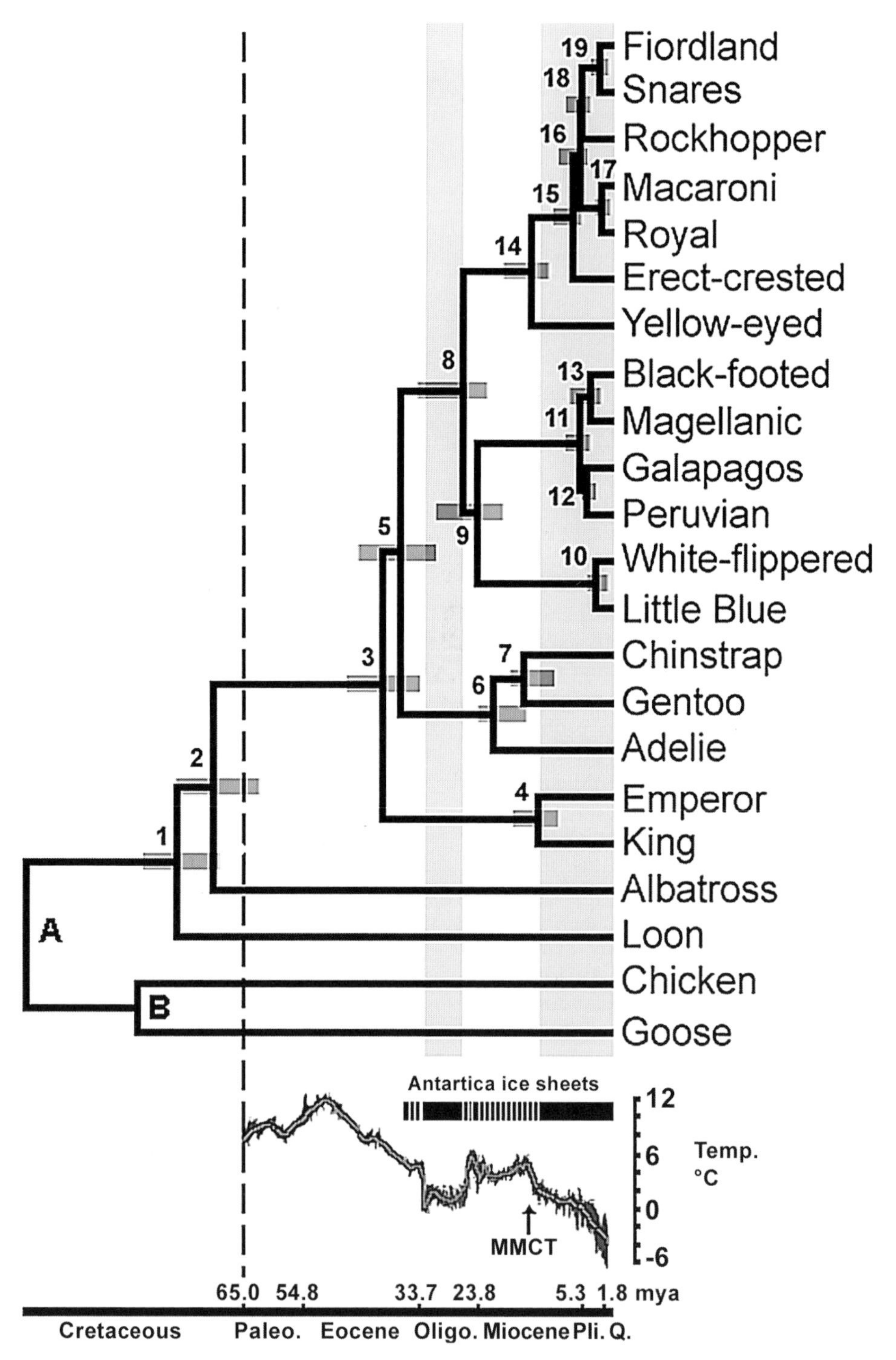


FIG. 5. Chronogram of penguin diversification from Baker et al. (2006). Nodes A and B were fixed at 104 and 90 mya. Gray bars at numbered internal nodes indicate credibility intervals (95%). Vertical dashed line indicates the Cretaceous–Tertiary boundary. Periods when Antarctica was ice-covered are projected as shaded gray rectangles in the chronogram. Ocean temperature is based on high-resolution deep-sea oxygen isotope records. The Middle Miocene Climate Transition (MMCT) is indicated by an arrow.

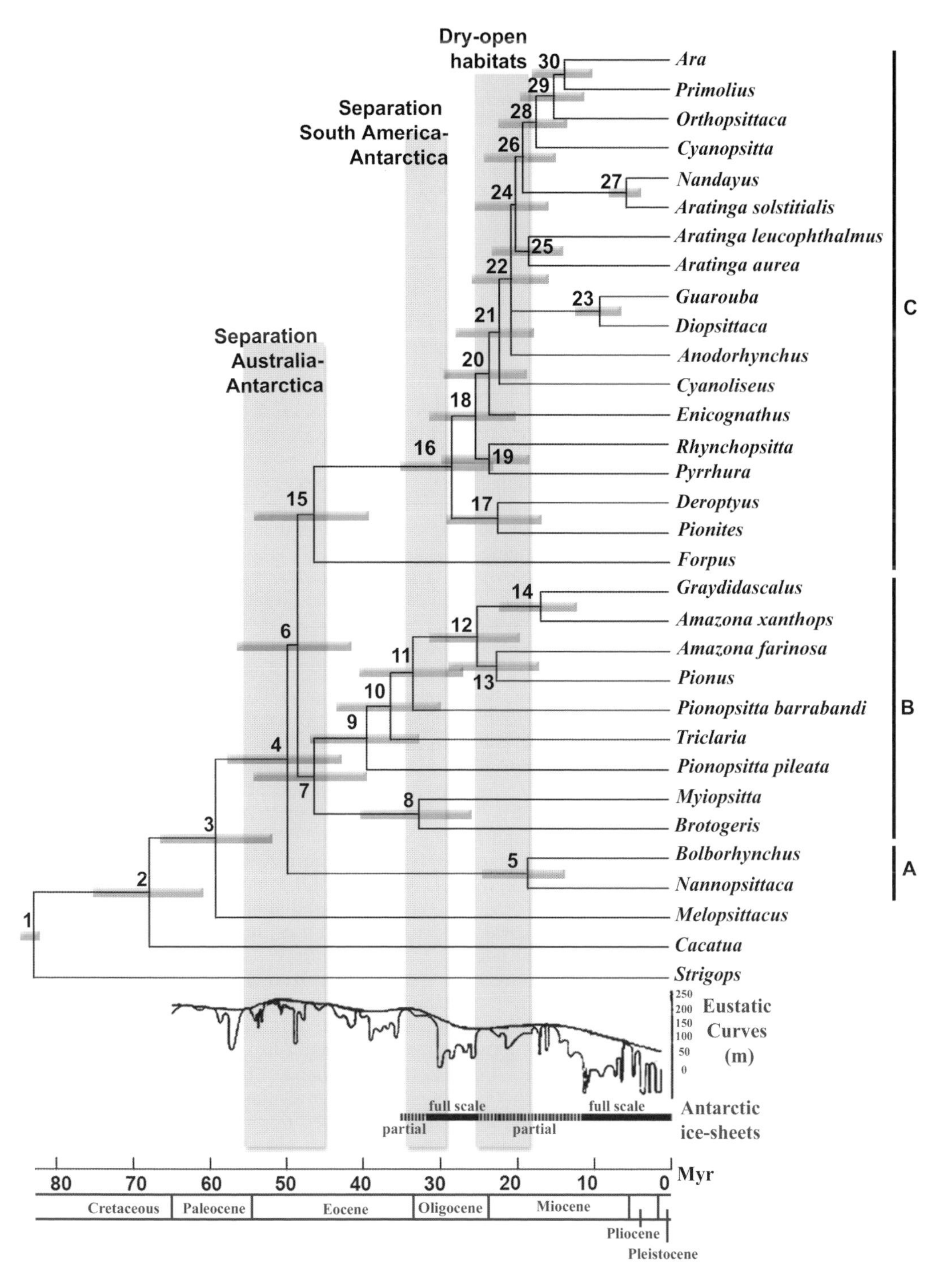


FIG. 6. Chronogram showing divergence times among the parrot genera and paleoevents possibly related to the Neotropical diversification. Numbers on nodes correspond to the estimated divergence times of lineages as given in table 3 of Tavares et al. (2006). Horizontal bars at nodes are 95% credibility intervals of divergence times. Clades are defined as A, parrotlets; B, amazons and allies; and C, macaws, conures, and allies. Eustatic curves of sea level and the extent of the Antarctic ice sheets are shown below the chronogram.

ACKNOWLEDGMENTS

I thank M. Burbidge and A. Given for permission to feature some of their unpublished Ph.D. research and the Natural Sciences and Engineering Research Council of Canada for financial support of my work over the years. I am grateful to S. Edwards and an anonymous reviewer for their constructive criticisms of an earlier draft of the manuscript.

LITERATURE CITED

AVISE, J. C. 2000. Phylogeography: The History and Formation of Species. Harvard University Press, Cambridge, Massachusetts.

BAKER, A. J., L. HUYNEN, O. HADDRATH, C. D. MILLAR, AND D. M. LAMBERT. 2005. Reconstructing the tempo and mode of evolution in an extinct clade of birds with ancient DNA: The giant moas of New Zealand. Proceedings of the National Academy of Sciences USA 102:8257–8262.

BAKER, A. J., AND A. MOEED. 1987. Rapid genetic differentiation and founder effect in colonizing populations of the Common Myna (*Acridotheres tristis*). Evolution 41:525–538.

BAKER, A. J., S. L. PEREIRA, O. HADDRATH, AND K. EDGE. 2006. Multiple gene evidence for expansion of extant penguins out of Antarctica in response to global cooling. Proceedings of the Royal Society of London, Series B 273:11–17.

BALLARD, J. W. O., AND M. C. WHITLOCK. 2004. The incomplete natural history of mitochondria. Molecular Ecology 13:729–744.

BARROWCLOUGH, G. F., N. K. JOHNSON, AND R. M. ZINK. 1985. On the nature of genic variation in birds. Pages 135–154 *in* Current Ornithology, vol. 2 (R. F. Johnston, Ed.). Plenum Press, New York.

BAZIN, E., S. GLEMIN, AND N. GALTIER. 2006. Population size does not alter mitochondrial genetic diversity in animals. Science 312: 570–572.

BENSCH, S., D. IRWIN, J. IRWIN, L. KVIST, AND S. ÅKESSON. 2006. Conflicting patterns of mitochondrial and nuclear DNA diversity in *Phylloscopus* warblers. Molecular Ecology 15: 161–171.

BRITO, P. H. 2005. The influence of Pleistocene glacial refugia on Tawny Owl genetic diversity and phylogeography in western Europe. Molecular Ecology 14:3077–3094.

BUNCE, M., T. H. WORTHY, T. FORD, W. HOPPITT, E. WILLERSLEV, A. DRUMMOND, AND A. COOPER. 2003. Extreme reversed sexual size dimorphism in the extinct New Zealand moa *Dinornis*. Nature 425:172–175.

BURBIDGE, M. L., R. M. COLBOURNE, H. A. ROBERTSON, AND A. J. BAKER. 2003. Molecular and other biological evidence supports the recognition of at least three species of brown kiwi. Conservation Genetics 4:167–177.

CICERO, C., AND N. K. JOHNSON. 1998. Molecular phylogeny and ecological diversification in a clade of New World songbirds (genus *Vireo*). Molecular Ecology 7:1359–1370.

CLEGG, S. M., S. D. DEGNAN, J. KIKKAWA, C. MORITZ, A. ESTOUP, AND I. P. F. OWENS. 2002. Genetic consequences of sequential founder events by an island-colonizing bird. Proceedings of the National Academy of Sciences USA 99: 8127–8132.

CORANDER, J., P. WALDMAN, AND J. SILLANPAA. 2003. Bayesian analysis of genetic differentiation among populations. Genetics 163:367–374.

EDWARDS, S. V., S. B. KINGAN, J. D. CALKINS, C. N. BALAKRISHNAN, W. B. JENNINGS, W. J. SWANSON, AND M. D. SORENSON. 2005. Speciation in birds: Genes, geography and sexual selection. Proceedings of the National Academy of Sciences USA 102:6550–6557.

GEMMELL, N. J. M., K. SCHWARTZ, AND B. C. ROBERTSON. 2004. Moa were many. Proceedings of the Royal Society of London, Series B 271: S430–S432.

GIVEN, A. D. 2004. Phylogenetics and population genetics of the Australasian Silver Gull (*Larus novaehollandiae*). Ph.D. dissertation, University of Toronto, Toronto, Ontario.

GREEN, R. E., J. KRAUSE, S. E. PTAK, A. W. BRIGGS, M. T. RONAN, J. F. SIMONS, L. DU, M. EGHOLM, J. K. ROTHBERG, M. PAUNOVIC, AND S. PAABO. 2006. Analysis of one million base pairs of Neanderthal DNA. Nature 444:330–336.

GRISWOLD, C., AND A. J. BAKER. 2002. Time to the most recent common ancestor and divergence times of populations of Common Chaffinches (*Fringilla coelebs*) in Europe and North Africa: Insights into Pleistocene refugia and postglacial colonizations. Evolution 56:143–153.

HADDRATH, O., AND A. J. BAKER. 2001. Complete mitochondrial DNA genome sequences of extinct birds: Ratite phylogenetics and the vicariance biogeography hypothesis. Proceedings of the Royal Society of London, Series B 268:1–7.

HEBERT, P. D. N., M. Y. STOECKLE, T. S. ZEMLAK, AND C. FRANCIS. 2004. Identification of birds through DNA barcodes. PLoS Biology 2: 1657–1663.

HEIN, J., M. SCHIERUP, AND C. WIUF. 2004. Gene Genealogies, Variation and Evolution: A Primer in Coalescent Theory. Oxford University Press, Oxford, United Kingdom.

HEY, J., AND R. NIELSEN. 2004. Multilocus methods for estimating population sizes, migration rates and divergence time, with applications to the divergence of *Drosophila pseudoobscura* and *D. persimilis*. Genetics 167:747–760.

Hickerson, M. J., C. P. Meyer, and C. Moritz. 2006. DNA barcoding will often fail to discover new animal species over broad parameter space. Systematic Biology 55:729–739.

Huynen, L., C. D. Millar, R. P. Scofield, and D. M. Lambert. 2003. Nuclear DNA detects species limits in ancient moa. Nature 425:175–178.

Irwin, D. E., S. Bensch, and T. D. Price. 2001. Speciation in a ring. Nature 409:333–337.

Jennings, W. B., and S. V. Edwards. 2005. Speciational history of Australian grass finches (*Poephila*) inferred from 30 gene trees. Evolution 59:2033–2047.

Johnson, N. K. 1980. Character variation and evolution of sibling species in the *Empidonax difficilis–flavescens* complex (Aves: Tyrannidae). University of California Publications in Zoology, no. 112.

Johnson, N. K., and C. Cicero. 2002. The role of ecologic diversification in sibling speciation of *Empidonax* flycatchers (Tyrannidae): Multigene evidence from mtDNA. Molecular Ecology 11: 2065–2081.

Johnson, N. K., J. V. Remsen, Jr., and C. Cicero. 1999. Resolution of the debate over species concepts in ornithology: A new comprehensive biologic species concept. Pages 1470–1482 *in* Acta XXII Congressus Internationalis Ornithologici (N. J. Adams and R. H. Slotow, Eds.). BirdLife South Africa, Johannesburg.

Lambert, D. M., A. Baker, L. Huynen, O. Haddrath, P. D. N. Hebert, and C. D. Millar. 2005. Is a large-scale DNA-based inventory of ancient life possible? Journal of Heredity 96:279–284.

Meier, R., K. Shiwang, G. Vaidya, and P. K. L. Ng. 2006. DNA barcoding and taxonomy in Diptera: A tale of high intraspecific variability and low identification success. Systematic Biology 55:715–728.

Moritz, C., and C. Cicero. 2004. DNA barcoding: Promise and pitfalls. PLoS Biology 2: 1529–1531.

Nielsen, R., and J. Wakeley. 2001. Distinguishing migration from isolation: A Markov chain Monte Carlo approach. Genetics 158:885–896.

Paton, T., O. Haddrath, and A. J. Baker. 2002. Complete mtDNA genome sequences show that modern birds are not descended from transitional shorebirds. Proceedings of the Royal Society of London, Series B 269:839–846.

Pereira, S. L., and A. J. Baker. 2006. A mitogenomic timescale for birds detects variable phylogenetic rates of molecular evolution and refutes the standard molecular clock. Molecular Biology and Evolution 23:1731–1740.

Poinar, H., C. Schwarz, J. Qi, B. Shapiro, R. D. E. MacPhee, B. Buigues, A. Tikhonov, D. H. Huson, L. P. Tomsho, A. Auch, M. Rampp, W. Miller, and S. C. Schuster. 2006. Metagenomics to paleogenomics: Large-scale sequencing of mammoth DNA. Science 311:392–394.

Pritchard, J. K., M. Stephens, and P. Donnelly. 2000. Inference of population structure using multilocus genotype data. Genetics 155:945–959.

Ronquist, F. R., and J. P. H. Huelsenbeck. 2003. MRBAYES: Bayesian inference of phylogeny. Bioinformatics 19:1572–1574.

Rozas, J., J. C. Sánchez-DelBarrio, X. Messeguer, and R. Rozas. 2003. DnaSP, DNA polymorphism analyses by the coalescent and other methods. Bioinformatics 19:2496–2497.

Sæther, B.-E., R. Lande, S. Engen, H. Weimerskirch, M. Lillegård, R. Altwegg, P. H. Becker, T. Bregnballe, J. E. Brommer, R. H. McCleery, and others. 2005. Generation time and temporal scaling of bird population dynamics. Nature 436:99–102.

Sorenson, M. D., K. M. Sefc, and R. B. Payne. 2003. Speciation by host switch in brood parasitic indigobirds. Nature 424:928–931.

Tavares, E. S., A. J. Baker, S. L. Pereira, and C. Y. Miyaki. 2006. Phylogenetic relationships and historical biogeography of Neotropical parrots (Psittaciformes: Psittacidae: Arini) inferred from mitochondrial and nuclear DNA sequences. Systematic Biology 55:454–470.

Thorne, J. L., and H. Kishino. 2002. Divergence time and evolutionary rate estimation with multilocus DNA data. Systematic Biology 51: 689–702.

Turbott, E. G. 1990. Checklist of the Birds of New Zealand and the Ross Dependency, Antarctica. Random Century, Auckland, New Zealand.

Wang, Z., A. J. Baker, G. Hill, and S. V. Edwards. 2003. Reconciling actual and inferred population histories in the House Finch (*Carpodacus mexicanus*) by AFLP analysis. Evolution 57: 2852–2864.

Wenink, P. W., A. J. Baker, H. Rosner, and M. G. J. Tilanus. 1996. Global mitochondrial phylogeography of Holarctic breeding Dunlins (*Calidris alpina*). Evolution 50:318–330.

Zink, R. M. 1997. Phylogeographic studies of North American birds. Pages 301–324 *in* Avian Molecular Evolution and Systematics (D. P. Mindell, Ed.). Academic Press, San Diego, California.

Zink, R. M. 2004. The role of subspecies in obscuring biological diversity and misleading conservation policy. Proceedings of the Royal Society London, Series B 271:561–564.

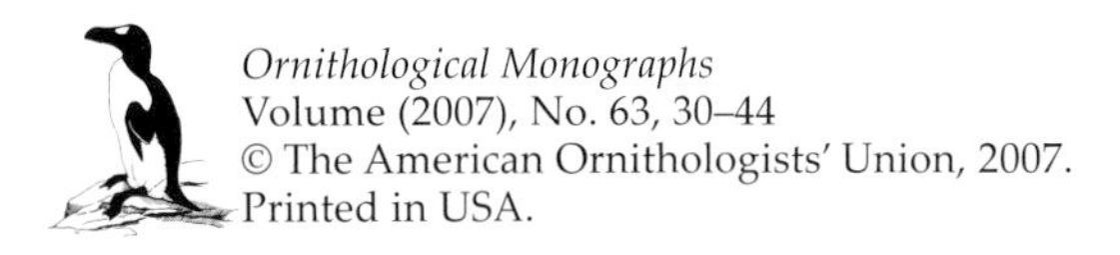
*Ornithological Monographs*
Volume (2007), No. 63, 30–44


CHAPTER 3

# VAINLY BEATING THE AIR: SPECIES-CONCEPT DEBATES NEED NOT IMPEDE PROGRESS IN SCIENCE OR CONSERVATION

Kevin Winker,[1,3] Deborah A. Rocque,[1,2] Thomas M. Braile,[1] and Christin L. Pruett[1]

[1]*University of Alaska Museum, 907 Yukon Drive, Fairbanks, Alaska 99775, USA; and*
[2]*U.S. Department of Interior, Fish and Wildlife Service, 1011 Tudor Road, Anchorage, Alaska 99503, USA*

Abstract.—Debate over species concepts has been a persistent theme in biology. We briefly summarize competing species concepts and facets of the debate itself. We maintain that the inherent subjectivity within all species concepts is likely to ensure continued disagreement on where to place species limits. Although the debate itself contributes to the understanding of speciation and evolutionary processes, it can take on political overtones through posturing, caricatures, and provocative statements. Empirically, neither basic nor applied science would seem to have been slowed appreciably because the species-concept debate remains unresolved. Similarly, continued disagreement must be placed in its proper context (e.g., be shelved) when considering the preservation of biodiversity. To a considerable extent, this has occurred within the conservation community. The biological species concept (BSC) and its inclusion of diagnosably distinct populations as subspecies remain dominant in ornithology. This may be attributable, in part, to the seemingly infinitely fine divisions possible under phylogenetic species concepts (PSC)—which, among other things, could strain public credulity over what constitutes a species. Nevertheless, the strengths of each of these concepts are being applied to improve our understanding of biodiversity. The longstanding disagreement over species concepts should not become an impediment to responsible conservation and wildlife management. It probably has not occurred broadly yet, but there may be potential for such an effect in the political arena. *Received 30 July 2006, accepted 23 February 2007.*

Resumen.—El concepto de especie es un tema polémico en biología. Resumimos brevemente los distintos conceptos de especie y las facetas del debate. Consideramos que la subjetividad inherente a todos los conceptos de especie garantiza un continuo desacuerdo en como delimitar las especies. Aunque el debate en si mismo contribuye a entender el proceso de especiación y otros procesos evolutivos, muchas veces se desarrolla en términos demasiado caricaturescos, burlescos y provocativos. De hecho, la continuidad del debate no parece haber ralentizado ni la ciencia básica ni la aplicada. De igual modo, el continuo desacuerdo debe contextualizarse cuando consideramos la conservación de la diversidad. El concepto biológico de especie (BSC) y la inclusión de las poblaciones diagnosticables como subespecies, sigue predominando en ornitología. Esto puede deberse en parte a las infinitas divisiones que se pueden realizar bajo el concepto filogenético de especie (PSC), lo que entre otras cosas, puede afectar el entendimiento del público en general sobre qué es una especie. Sin embargo, los puntos fuertes de cada concepto se están aplicando para llegar a un mejor entendimiento de la biodiversidad. Este desacuerdo histórico sobre el concepto de especie no tiene por qué ser un impedimento para una política responsable para la conservación y gestión de la vida salvaje.

---

[3]E-mail: ffksw@uaf.edu

> ...Nor shall I here discuss the various definitions which have been given of the term species. No one definition has as yet satisfied all naturalists; yet every naturalist knows vaguely what he means when he speaks of a species. (Darwin 1859:44)

> ...But to discuss whether they are rightly called species or varieties, before any definition of these terms has been generally accepted, is vainly to beat the air. (Darwin 1859:49)

What defines a species? This question has divided biologists and been a persistent theme in biology for >150 years (e.g., Darwin 1859; Mayr 1942, 1993; Wiley 1978; Cracraft 1983; Paterson 1985; Templeton 1989; Nixon and Wheeler 1990; Mallet 1995; Avise and Wollenberg 1997; de Quieroz 1998; Harrison 1998; Johnson et al. 1999; Wheeler and Meier 2000; Hey 2001; Coyne and Orr 2004). The debate over species concepts can evoke such passion that arguments often focus on the weaknesses of competing concepts and rarely acknowledge the benefits of the debate itself. Given that we are attempting to place broadly definitive lines upon a continuous process, the debate is certain to continue. At present, no species concept can claim convincingly to provide rigorous methods for objectively defining when speciation has occurred—at least not in a manner that can be usefully applied within or across taxa on a broad scale. Nonetheless, the lack of a unifying species concept has not hindered serious progress in the biological sciences or in the development of appropriate units for conservation priorities.

We find humor in Darwin's statement that defining species is like "vainly beating the air" (Darwin 1859:49), but the ongoing debate has helped to point out areas where further investigation is needed. We propose that taxonomists and systematists put aside arguments for scrapping the current species structure until a more robust and objective alternative can be implemented with broad agreement. To date, we know of no such alternative. Here, we briefly review the species-concept debate and then address several questions: (1) Has the biological species concept (BSC) failed us? (2) Is the phylogenetic species concept (PSC), widely considered the preferred alternative to the BSC in ornithology, an objective and robust concept ready to impose new species boundaries? (3) Is it appropriate to abandon the BSC for promotion of conservation initiatives? And (4) has disagreement about species concepts impeded basic or applied science? Finally, we consider the realities of geographically partitioned variation within species and the scientific and legal (at least in the United States and Canada) bases for continuing to recognize divisions below the species level.

## Process-based Species Concepts

Several species concepts focus on reproductive mechanisms or the processes that define species. The concepts receiving most attention in this category include the BSC (Mayr 1942), the recognition species concept (RSC; Paterson 1985), and the reproductive-cohesion species concept (CSC; Templeton 1989). In what is widely referred to as the BSC, Mayr (1963: 19) defined species as "groups of actually or potentially interbreeding natural populations which are reproductively isolated from other such groups." The BSC, and our working model of it, is the most widely accepted definition of species.

Despite its widespread use, the BSC has been criticized for failing to identify species in allopatric populations, for its difficulty in dealing with hybridization, and for its inapplicability to asexually reproducing organisms. Paterson (1985) and Templeton (1989) recognized the usefulness of defining species by isolating mechanisms and attempted to improve the BSC by adding additional criteria for naming species. The RSC emphasizes the evolutionary development of prezygotic isolating mechanisms, defining species as "the most inclusive population of individual biparental organisms which share a common fertilization system" (Paterson 1985:15). The RSC changed the focus in defining species from mechanisms that prevent mating to those that facilitate reproduction. Under the RSC, specific mate-recognition systems serve as signal-and-response interactions between individuals of a species (Paterson 1985) and are the main criteria by which species are defined. Because mate recognition can occur only within a species, hybridization is theoretically impossible under the RSC (King 1993). Although Paterson attempted to improve the BSC by identifying species in hybrid zones, the RSC has received criticism for difficulty in evaluating specific mate-recognition systems and its inability to deal with asexual or allopatric populations (Templeton 1989, Mayden 1997).

Templeton (1989) introduced the CSC, which focuses on processes that maintain similarity among populations. Under the CSC, species are "the most inclusive population of individuals having the potential for phenotypic cohesion through intrinsic cohesion mechanisms" (Templeton 1989:12). Rather than focusing on the mechanisms that separate species (i.e., BSC and RSC), the CSC focuses on factors, both genetic and phenotypic, that hold a species together. The CSC merges central ideas from process- and pattern-based concepts into a species concept that is operational for both sexual and asexual organisms. Although the CSC is operational for a wide variety of taxa, it has remained unimplemented, mainly because "phenotypic cohesion" remains largely undefined (Endler 1989). Templeton's effort to incorporate pattern into a species concept was not new; most of the major competing concepts to the BSC are pattern-based.

### Pattern-based Species Concepts

Pattern-based species concepts focus on evolutionary history, phylogenetic relationships, and character states. These concepts include the evolutionary species concept (ESC; Wiley 1978), the PSC (Cracraft 1983), the genealogical species concept (GSC; Baum and Shaw 1995), and the genotypic-cluster species definition (Mallet 1995). The ESC, with its roots in paleontology, defines a species as "a single lineage of ancestor—descendant populations which maintains its identity from other such lineages and which has its own evolutionary tendencies and historical fate" (Wiley 1978:18). Because the ESC includes ancestral populations, it is difficult to test scientifically and does not seem to be applied much outside the field of paleontology, although others have shown interest in it (e.g., Peterson 1998, Taylor et al. 2005).

The PSC has received the most attention and consideration recently as a potential alternative to the BSC (e.g., Cracraft 1989, 1997; Nixon and Wheeler 1990; Zink and McKitrick 1995), and in ornithology (at least) it is routinely portrayed as the leading opponent to the BSC. There are actually multiple PSCs, but they all share a focus on identification of historically related groups (Coyne and Orr 2004). Here, we focus on the most ornithologically relevant aspects and consider the PSC in the singular. Cracraft (1997:329) defined a species under the PSC as "the smallest population or group of populations within which there is a parental pattern of ancestry and descent and which is diagnosable by unique combinations of character-states." This definition was modified from its original form to address earlier criticisms.

The PSC has been criticized by other proponents of pattern-based concepts for relying too heavily on the principles of cladistics and diagnosing species in terms of apomorphies (characters derived and different from the ancestral condition; Nelson and Platnick 1980, Mallet 1995). A general fear among many of these critics is that perfectly good species that lack such apomorphies will not be diagnosed properly. To avoid the use of apomorphies as the unit of diagnosability, Mallet (1995:296) proposed the genotypic-cluster species definition, which defines species as

> distinguishable groups of individuals that have few or no intermediates when in contact...[C]lusters are recognized by a deficit of intermediates, both at single loci (heterozygote deficits) and at multiple loci (strong correlations or disequilibria between loci that are divergent between clusters).

Although the genotypic-cluster definition can diagnose species without apomorphies, it confounds species diagnosis with the requirement that species must be "in contact" (e.g., parapatric or sympatric; Harrison 1998).

Baum and Shaw (1995) also chose to reduce the emphasis placed on character states or apomorphies by the PSC, proposing the GSC (considered by Coyne and Orr [2004] to be a third general version of the PSC). The GSC defines species as "exclusive groups of organisms, where an exclusive group is one whose members are more closely related to each other than any other organisms outside the group" (Baum and Shaw 1995:290). Relatedness under the GSC is determined by the concordance of gene genealogies; however, the amount of concordance necessary between two populations for them to be named genealogical species is not specified (i.e., all gene genealogies or just some; Harrison 1998). The amount of introgression between species and across a species' genome can vary; consequently, alleles can become fixed at some loci and not at others, resulting in discordant gene genealogies (Harrison 1998). Under these circumstances, the GSC, which requires concordance over a number of loci, may be too stringent in how it defines species limits or boundaries (Harrison 1998).

Finally, with the advent of genetic tools, we are seeing a resurgence in typological species concepts, though proponents are not naming their concepts as such. Typological species concepts based on phenotype have been disfavored since the mid-19th century because of inherent conceptual flaws (Mayr 1963, 1982). Mayr (1963: 5–6) wrote:

> The replacement of typological thinking by population thinking is perhaps the greatest conceptual revolution that has taken place in biology.... Virtually every major controversy in the field of evolution has been between a typologist and a populationist.

Typological species are defined by specifying an arbitrary degree of morphological or genetic difference (Coyne and Orr 2004). Naming groups "species" on the basis of an arbitrary degree of genetic divergence seems to have become widespread (e.g., König et al. 1999, Hebert et al. 2004). Although it is true that genetic divergence generally increases as lineages diverge, there is no genetic measure yet that enables us to determine when speciation has occurred between lineages in the class Aves (e.g., Funk and Omland 2003, Johnson and Cicero 2004, Moritz and Cicero 2004). Genetic diversity is certainly part of biodiversity, but using simplistic genetic distance data to designate species status is decidedly a step backwards in our efforts to understand biodiversity and its generation—in every way a return to the flaws of historical typological species concepts (see Mayr 1963, 1982). Given the stochastic and often selectively neutral processes inherent in DNA mutation and evolution, a lineage showing reciprocal monophyly in a single locus (to some researchers an indication of a phylogenetic species) may represent another of these arbitrary thresholds (see also Avise 2000b, Coyne and Orr 2004). At the least, such approaches do not accurately reflect what we know about the genetics of speciation (Orr 2001; Avise 2000b, 2004; Coyne and Orr 2004).

## Should We Abandon the Biological Species Concept?

No—although it has its problems. The BSC does not apply to asexual populations, because it is based on reproductive properties. Mayr (1996: 266) was unapologetic for this flaw, believing that "a species definition that is equally applicable to both sexually reproducing and asexual populations misses the basic characteristics of the biological species definition," and he turned to the specialists to sort out species in these organisms. That no single species concept can be applied equally well to all taxa has been stated repeatedly (e.g., Bock 1992, Mayr 1996).

The BSC's requirement of reproductive isolation creates confusion in naming allopatric species because this requirement is generally not tested. This is commonly viewed as the concept's biggest drawback (e.g., Zink 1996). Mayr's original definition did not include "potentially" with regard to interbreeding natural populations; it was inserted later with the hope that it would diminish the criticism regarding allopatric populations. For some critics, it did not. In areas of allopatry, determinations of reproductive isolating mechanisms are often extrapolated using cues from morphology, behavior, and ecology (Cracraft 1989, Johnson et al. 1999, Helbig et al. 2002) and require a certain degree of subjectivity. In this sense, the BSC is guilty of inferring process from pattern (Harrison 1998). The processes of differentiation create patterns, and making inferences from the latter has been a standard practice in BSC-based taxonomy.

Expounding on classic methods regarding relative differentiation between forms, Mayr (1969:196–97) suggested that the status of reproductive isolation in allopatric populations be inferred by using phenotypic divergence as a yardstick, employing three types of divergence data in closely allied taxa "to calibrate such a scale": differences between sympatric species, between intergrading subspecies within species, and between hybridizing populations. Such methods are hardly foolproof (Mayr 1969), but this approach remains important today (see Remsen 2005). Experimental approaches to determining whether reproductive isolation exists between allopatric populations are possible (e.g., through playback, captive breeding, or mate-choice experiments), but interpretation of results is confounded by many factors (Mayr

1969), not the least of which is the difficulty of determining the fitness of hybrid individuals in relation to pure individuals of the parent lineages.

Known hybridization of distinct BSC species has been proposed as another major limitation of this concept (Cracraft 1997). Although ~10% of bird species can hybridize (Grant and Grant 1992), the extent of this problem is limited because much hybridization does not result in production of fertile offspring or in individuals with increased fitness likely to drive evolutionary change. Also, effective gene flow and its effects are considered important by the working model of the BSC; the extent and nature of hybridization can be studied and incorporated within this concept (e.g., Johnson et al. 1999). Basically, the failure of prezygotic isolating mechanisms to prevent hybridization still leaves the opportunity for postzygotic isolating mechanisms such as reduced fertility or fitness to operate to prevent the likelihood of lineage reticulation. Additionally, such lineage crossings are likely a challenge to any species concept when they make it difficult to recover evolutionary history (Grant and Grant 1992).

Mace and Collar (2002) did not consider a movement to a phylogenetic species concept useful from the conservation perspective, but they recognized that a needed "unlumping" of biological species of birds was given impetus by PSC advocates in their criticisms of the BSC. They, like Avise (2000a), considered that the best way forward was through accommodation (Mace and Collar 2002). This accommodation is probably coming to pass, at least in part (see below). Despite many criticisms, the BSC remains the preferred species concept in well-studied groups such as birds (Avise and Wollenberg 1997, Johnson et al. 1999, Avise 2004). Although its many strengths have been widely recognized, perhaps the most important of these is its inherent inclusion of population thinking (see, e.g., Avise 2000a, Coyne and Orr 2004).

### Should We Implement the Phylogenetic Species Concept to Define New Species Boundaries?

The ability to name asexual organisms is an advantage of the PSC, but by emphasizing pattern the PSC has the potential to ignore processes important to speciation in sexually reproducing organisms. Proponents of the PSC may recognize the importance of process (interbreeding, reproductive isolation, and barriers to genetic exchange), but process is not among the criteria used to define phylogenetic species (Harrison 1998). We know, however, that gene flow is of profound importance in the speciation process.

Another criticism of the PSC concerns its emphasis on cladistics and the identification of apomorphies. Advocates of the PSC acknowledge this concern but remark that all species are hypotheses and that designations may change with the advent of new information (Cracraft 1997). In this approach, PSC proponents assume that phenotypic diagnostic characters are genetically based. By relying on these identification techniques, the PSC confines the speciation process to a simple presence–absence condition.

This is apparent in the manner in which the PSC deals with reticulation (the evolutionary reuniting of differentiated lineages) and hybridization. The PSC has been accused of ignoring reticulation and the existence of paraphyly (a species derived from a common ancestor but which does not include all the descendants of that ancestor), and the single-locus approach of mitochondrial DNA (mtDNA) phylogenetic studies and determination of species limits can clearly run afoul of polyphyly (a species derived from two or more ancestors) and paraphyly (Patton and Smith 1989, Avise and Wollenberg 1997, Johnson et al. 1999, Funk and Omland 2003). The BSC is able to group nonhistorical taxa as species when those taxa come into contact and freely interbreed (e.g., Northern Flicker, *Colaptes auratus*; Moore et al. 1991). The PSC, by focusing on historical pattern, could diagnose freely hybridizing non-sister taxa as separate species; the PSC simply ignores gene flow between them. Advocates of the PSC maintain that hybridization is a useless criterion in delineating species (Cracraft 1997). But in attacking the BSC on hybridization, PSC advocates tend to view the ability to interbreed as a single "character" rather than as the complex, variable process for genetic interchange that it is. In ornithology, the working model of the BSC does not treat hybridization as a presence–absence "character state," but rather assesses its likely importance by its degree and probable evolutionary consequences. This working model of the BSC ignores hybridization between lineages that is an evolutionary dead end (e.g., compare Gray 1958 and American Ornithologists' Union [AOU] 1998; see also Johnson et al. 1999).

It has been argued that the PSC is not applicable in its current form because of difficulties in implementing species diagnoses (Avise and Wollenberg 1997; see also Avise 2000b). Critics suggest that using detailed morphological analyses and modern molecular evolutionary techniques will lead to a proliferation of the number of species (Mayr 1993) and ultimately to the ability to diagnose individuals (Avise and Ball 1990; Avise 2000b, 2004; Mallet 1995). The PSC provides no definition of "diagnosable character states," leading to subjectivity in recognizing species. This subjectivity has the potential to create an imbalance between taxa (King 1993), creating new species in well-studied organisms while cryptic species in less-studied organisms may go undetected. Thus, as Collar (1997) pointed out, the PSC may produce species that are not the smallest diagnosable clusters, but rather those diagnosable to the point at which the search was abandoned. Johnson et al. (1999) presented evidence that this has been done already. Indeed, molecular techniques are so robust in diagnosing lineages to the individual level that Avise (2004:361) stated that "most individuals and family units within sexually reproducing species can be distinguished from one another with high-resolution molecular assays." Clearly, diagnosability under such modern methods requires some subjectively placed thresholds to associate diagnosed units with species limits.

Several species concepts have been proposed under process- and pattern-based ideologies, and their introduction has contributed to the understanding of speciation and the evolutionary process. Insofar as species concepts have been forged and reforged to aid progress in understanding biodiversity and evolutionary processes, the absence of agreement on a species concept has probably not been a major impediment to scientific progress. It even seems likely that some principles of competing concepts have been helpfully incorporated into taxonomy in cases that are difficult to resolve under any concept. But a fervent debate continues, especially between proponents of the BSC and the PSC, with each side extolling the virtues of one concept and emphasizing the flaws of the other. The PSC, in particular, has generated heated discussions of species limits, but it has little likelihood of being adopted wholly (see, e.g., Avise 2000b). We believe that phylogenetic research can help us to define species boundaries, develop taxonomic hierarchies, infer evolutionary relationships, and help prioritize conservation. However, we find that the PSC is no less subjective in defining species limits than the BSC; it simply moves the subjectivity to another dimension of the problem (see also Avise and Wollenberg 1997, Johnson et al. 1999, Remsen 2005). Single-locus reconstructions of the evolutionary history of populations or species can be illuminating or deceiving regarding species units (e.g., Cronin 1993, Talbot and Shields 1996, Funk and Omland 2003, Avise 2004). In this regard, the simplistic totting up of largely neutral genetic changes between lineages (e.g., König et al. 1999, Hebert et al. 2004) should be shunned as another potential metric by which to determine species status; this setting of arbitrary thresholds, as noted above, is a throwback to typological thinking and its associated overly simplistic species concept.

## The Reality of Subspecies

Within zoology, ornithology has been perhaps the most stalwart bastion of the recognition of subspecies as formally named intraspecific taxonomic units (Cutright and Brodhead 1981), despite a nihilistic attitude toward this taxonomic entity by some PSC advocates (e.g., McKitrick and Zink 1988, Zink 2004). Subspecies—interbreeding (or likely capable of interbreeding) but phenotypically diagnosable populations—have long been considered by many to be a useful, albeit messy, taxonomic unit. Although subspecies have been neglected by the American Ornithologists' Union since the 5th edition of the *Check-list of North American Birds* (AOU 1957), they nevertheless remain an important aspect of described avian biodiversity (e.g., Peters et al. 1934–1986, del Hoyo et al. 1992–2005, Dickinson 2003).

Mitochondrial DNA sequence data indicate that conspecific populations can be structured at a wide variety of evolutionary depths (Avise 2000a). Although subspecies have fallen into disfavor with some because they comprise a mixed bag of units representing variable and often unknown levels of evolutionary differentiation, the PSC would elevate these units to full species status when diagnosable. This would, in effect, decrease the average level of differentiation among species while increasing the variation of among-species differentiation encompassed within a genus. Furthermore, it would ignore geographically partitioned variation below the species level if not fully diagnosable (e.g., through reciprocal monophyly of molecular markers). Although many subspecies under the BSC will prove to be perfectly good species with further study, changing species concepts to make it so seems rather extreme. In addition, to sweep remaining intraspecific variation under the proverbial rug by denying any formal taxonomic status for those subspecies not making species status (under whatever sort of revision or criteria) is, in our view, a disservice to historic studies and to modern understanding of avian diversity.

At intraspecific levels we expect the largely neutral genetic variation that dominates current genetic-sequence-based data sets to be decoupled from differentiation attributable to selection. Thus, mtDNA sequence data have little bearing on the validity of named subspecies, which are based on among-population differences in phenotype that are more likely than mtDNA differences to be the result of selection (e.g., Mumme et al. 2006). However, subspecific variation provides a suggestion of underlying genetic differentiation, and genetic data can help us understand some of the evolutionary history of intraspecific variation. Genotypic data also can be valuable for genetic diagnoses of populations and regions that warrant special management or conservation importance—as long as we understand the nature of concordance and discordance between phenotypic and genotypic data sets.

An excellent example of subspecific variation that exhibits both morphological and genetic concordance among described subspecies (i.e., partitioned geographic variation correlated in both phenotype and genotype) occurs among Song Sparrow (*Melospiza melodia*) populations in northwestern North America. Genetic study of populations in this region showed a strong concordance between population-level microsatellite data and previously described, morphologically based subspecies (Fig. 1; Pruett and Winker 2005). These subspecific units do not meet anyone's definition of species, with gene flow occurring among most populations and a lack of reciprocal monophyly exhibited among traditionally used characters (Gibson and Kessel 1997, Pruett 2002, Pruett and Winker 2005). However, we believe that they clearly indicate the value to basic and applied science of formal taxonomic recognition of geographically partitioned variation. The BSC's inclusion of such subspecific units can be viewed as a decided advantage, but this does not validate the reality of all described subspecies. For example, the only other Song Sparrow subspecies (*M. m. amaka*) in this region that did not fit this pattern was found to be an undiagnosable and, thus, an invalid taxon after additional genetic and morphological study (Pruett et al. 2004; similar treatment without presentation of relevant data was provided by Gibson and Kessel [1997] and Arcese et al. [2002]).

Although some subspecies do not reflect biological reality (often having been based on insufficient evidence) and thus are virtually useless or even deceiving, others appear to be good indicators of how variation is distributed across a species' geographic range. Continued recognition of subspecies has several advantages. As Mayr (1969) pointed out, the trinomial informs us about the closest relationship and the allopatric condition of breeding populations. Subspecies are also taxonomic bookmarks, informing scientists and wildlife managers that a species is not homogeneous throughout its range. Finally, the trinomial can serve as a form of conservation triage, in that it relegates minor geographic variants to a lower rank (and thus a lower conservation priority) than the more distinct units that practically everyone would agree are species (J. V. Remsen, Jr. pers. comm.).

As Hey (2001) pointed out, evolutionary processes that created the patterns recognized today occurred largely in the past, and the place where these groups exist now is at the "wave front" of the evolutionary processes of the present. Too great a focus on events of the distant past risks a disregard for diversity generated in response to recent selection (see also Crandall et al. 2000); subspecific morphology may reflect more of the latter than the former. Debates about the meaning of such partitioned variation with regard to adaptation, natural selection, and environmental influences on development (phenotypic plasticity) are another issue that we will not consider here; the causes of subspecific variation represent a suite of questions

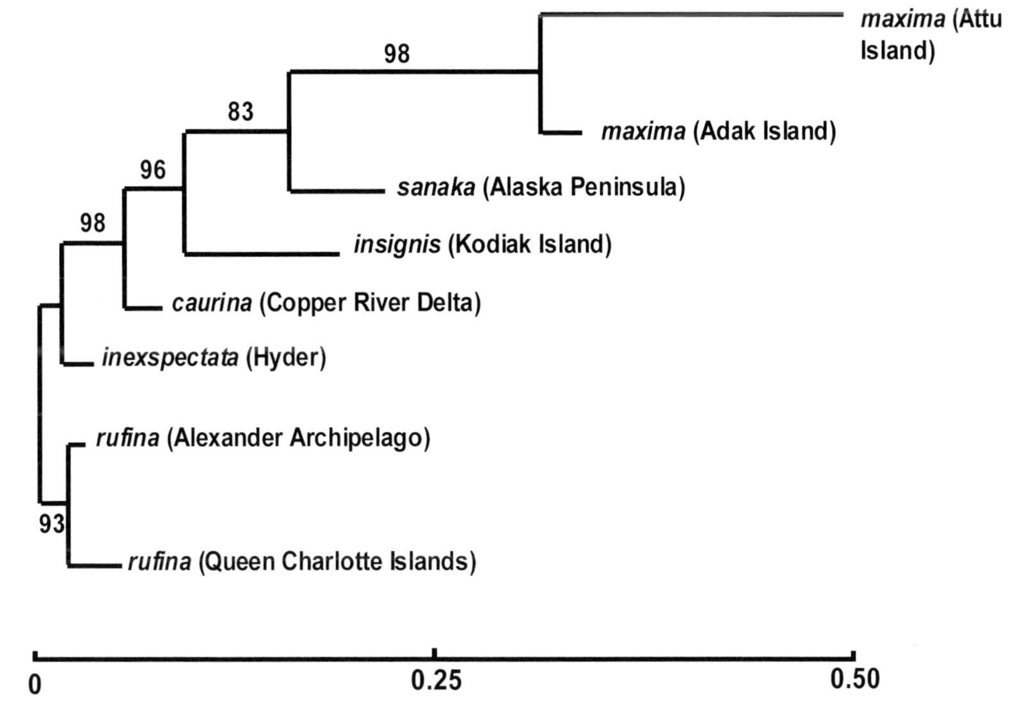


FIG. 1. Bootstrapped distance tree of Song Sparrow (*Melospiza melodia*) populations and subspecies across northwestern North America based on Nei's (1978) genetic distance using eight microsatellite loci. Genetic distance values are shown below the tree and bootstrap support for each branch is listed above that branch (adapted from Pruett and Winker 2005).

stimulated in large part because we recognize that such variation exists. Finally, as long as adaptation remains a possible reason for partitioned phenotypic variation, management and conservation of such evolutionary potential remains a legitimate goal.

## CONSERVATION

Conservation biology has the potential to be more affected by a change in species concept than any other discipline. How can we conserve what we cannot define? Realizing the effects that a species concept can have on preserving biodiversity, both sides of the BSC–PSC debate have looked to conservation biology to support their positions. For example, proponents of the PSC have claimed that "lumping" genetically diverse, evolutionarily distinct, and geographically separate populations into single species is detrimental to conservation efforts and have therefore recommended a change in concept (Zink and McKitrick 1995, Cracraft 1997). As conservation efforts move from a species- to a community-level focus, PSC advocates believe that an understanding of patterns of diversity and evolutionary history is essential (Zink and McKitrick 1995). Cracraft (1997) claimed that lower species limits under the PSC would give international governing agencies more legal incentive to protect these units. How such perennially resource-limited agencies would find the support to widen conservation efforts under a simple definition change, however, is unclear. Some conservation biologists believe that implementing the PSC would increase the total number of threatened species and devalue their status (e.g., Collar 1997). Conversely, BSC supporters charge that a change in species concepts is not only logistically impossible, but unnecessary because the Endangered Species Act (ESA) in the United States specifically includes protection of subspecies of vertebrates and plants and distinct population segments of vertebrates (ESA 1973). Setting conservation priorities for these subspecific units is not at all unusual (e.g., O'Brien and Mayr 1991), and Canada has recently adopted similar subspecific protections (e.g., Committee on the Status of Endangered Wildlife in Canada [COSEWIC] 2005).

Appropriate units for conservation attention can be and have been developed without assigning formal taxonomic status. The taxonomic "trick" of lowering species boundaries will not magically increase funding or concern for biological diversity in various national or international governing agencies. Although conservation biology can be considered a crisis discipline, taxonomy probably should not be (but see Godfray 2002, Mace 2004). We believe that lowering the bar for species recognition by changing the definition of what constitutes a species is a bit of professional legerdemain that will cause a portion of the nonprofessional public to lose faith in the biological sciences, especially in how we treat such widely visible groups as birds and other vertebrates. Applying the term "species" to vertebrate groups that are not discernible to the untrained observer (or even to professionals without genetic analyses, for example) is something to be strenuously avoided. More than 100 years ago, Teddy Roosevelt (before he was President of the United States) eloquently expressed this same barrier to acceptance of what we now term oversplitting:

> [Dr. Merriam] will do his work, if not in better shape, at least in a manner which will make it more readily understood by outsiders, if he proceeds on the theory that he is going to try to establish different species only when there are real fundamental differences, instead of cumbering up the books with hundreds of specific titles which will always be meaningless to any but a limited number of technical experts, and which, even to them, will often serve chiefly to obscure the relationships of the different animals by over-emphasis on minute points of variation. It is not a good thing to let the houses obscure the city. (Roosevelt 1897:880)

Despite the claims of various species-concept factions, conservation biology has flourished while this debate has continued (e.g., BirdLife International 2000, Norris and Pain 2002). Conservation units such as subspecies, evolutionarily significant units (ESUs), distinct population segments (DPSs), management units (MUs), and designatable units (DUs) have been developed to prioritize within-species diversity (ESA 1973; Ryder 1986; Moritz 1994, 1996; Green 2005; COSEWIC 2005). These units differ in their focus. For example, MUs represent sets of populations that are currently demographically independent, whereas ESUs are based on historically isolated sets of populations that together encompass the evolutionary diversity of a taxon (Moritz 1994). These subspecific units are not without their own inherent difficulties of definition (e.g., Paetkau 1999, Crandall et al. 2000), but they can provide a common, taxonomically largely neutral, ground—where, for example, the loaded term "species" can be ignored—upon which the majority of biologists can agree, regardless of species concepts. For example, Barrowclough and Flesness (1996) simply equated phylogenetic species with ESUs and discussed the latter to communicate more effectively with conservation biologists.

The collective job of conservation biologists, systematists, taxonomists, and wildlife managers is to recognize, conserve, and manage biological diversity from the population to the species level. These complex subjects of species concepts, geographic variation, differentiation, taxonomy, and hybridization play a central role in this business. But recognition of units, regardless of what we call them, can be accomplished. Assessment of these units' uniqueness and value are critical to future management and conservation, and they require the best scientific evidence available. Clearly, units for conservation priority can and have been developed and determined without major impediment from the species-concept debate and without scrapping current species concepts. Indeed, in looking forward, it seems that a BSC framework is more likely to prevail if only because it enhances stability (Agapow et al. 2004, Mace 2004). Regardless of species concepts and debates, however, we should recognize that an ever-growing body of data and an improved understanding of evolution are, in part, contributing to this effervescent front.

One area where the species-concept debate has potential to be detrimental is in the political arena. Issues regarding conservation of our planet's biodiversity are fertile battlegrounds for disagreement. Public and political support is germane to protecting populations and communities; conservation priorities are weighed against economic and political priorities. The public and politicians look to biologists to describe the organisms, populations, and higher units that need consideration. Public or institutional debates often ensue, and often during these debates the science is questioned. For example, scientific debates over concepts in evolutionary biology have left an opening for religious agendas in the U.S. education system (e.g., the "intelligent design" movement). Scientific debate over the causes and effects of global warming has left an opening for governments to delay development or implementation of effective policies to address the problem. It is not unreasonable to consider that debates over species concepts might provide an opening for government agencies or pro-development forces to neglect or postpone costly conservation efforts until species concepts are resolved. Biologists should recognize the social dimension to this scientific issue and be prepared to stand united on the fact that global biodiversity losses are—and will likely continue to be—staggering, and that responsible management and conservation of this diversity is imperative. Our longstanding disagreement over species concepts should not become an impediment to these goals. We should be prepared to shelve the debate and agree on the importance of responsible conservation and management regardless of what labels are used to denote the focal unit(s). Hey et al. (2003) drew parallels between species uncertainties and medical diagnoses or major weather events such as droughts—all are areas where professionals deal with inherent uncertainties in matters of great importance to society, and the uncertainties in each can be explained.

Conservation biology has moved forward in three major ways, despite disagreement over species concepts: (1) top-down conservation, working with higher-level taxa to formulate conservation plans, or effectively implementing conservation at the ecosystem level by preserving communities that exhibit shared patterns of evolutionary history (e.g., Moritz 1996, Williams et al. 1997, Villaseñor et al. 2005, Brooks et al. 2006); (2) "coat-tails" conservation, in which management or preservation of particularly desirable organisms (e.g., old-growth trees, cranes, waterfowl) necessarily brings additional biodiversity along; and (3) redefining and proceduralizing the recognition of units on the population-to-species continuum. This area has moved forward well since Mayr's (1969) *Principles of Systematic Zoology*, and recent advances can be seen in Helbig et al. (2002), Patten and Unitt (2002), and Sites and Marshall (2004). In addition, ESUs, MUs, DPSs, and DUs have been brought to bear in conservation and management. Mace's (2004) suggestion and the demonstrated utility (e.g., Bulgin et al. 2003, Pruett et al. 2004, Barrowclough et al. 2005) of combining taxonomy–systematics and conservation biology to address issues in the biopolitical realm is a likely roadmap to success. Likewise, Moritz's (2002) suggestion to aim for protection of both the patterns of biodiversity and the processes that generate this diversity emphasizes a most sensible multidimensional aspect to the scientific basis of biological conservation. Finally, we should remember that congruence among multiple data sets will enhance acceptance of determinations regardless of the units under consideration.

## Conclusions

Although there is little agreement on which species concept or suite of species concepts is best, we must acknowledge that the debate has helped guide speciation research and furthered our knowledge and understanding of evolutionary processes. Many advocates of a PSC have suggested complete abandonment of the BSC (e.g., Nixon and Wheeler 1990, Zink and McKitrick 1995, Cracraft 1997). In our opinion, this would be a grave mistake. In establishing and defending the BSC, biologists have contributed to our understanding of the process of speciation in sexually reproducing organisms, thus identifying mechanisms that lead to reproductive isolation. Similarly, extensive research on phylogenetic lineages under the framework of the PSC has increased our understanding of historical patterns of differentiation (Avise 2000a). Modern molecular evolutionary techniques are bridging the historical gap between population genetics and systematics, and both sides are beginning to realize how intimately related their fields are (Avise and Walker 1999, Avise 2000a). Undoubtedly, the strengths of each of these concepts are being applied to improve our understanding of biodiversity.

Cracraft (2000) did a service in pointing out the "my concept is best" phenomenon among the species-concept debaters. This strong and repeated trend in the debate emphasizes its political and subjective nature. Levin's (1979) view that species are essentially tools or abstract constructs that we create to handle biodiversity is a useful insight. Endler (1989) discussed the use of different concepts for different aspects of the study of speciation, further strengthening the analogy to tools. Hey (2001) determined that the species problem is inherent in the clash between the human propensity to categorize and the desire among biologists to make the categorical bin of "species" concordant with an evolutionary group. Regardless of how future facets of the species-concept debate develop, we can be assured that it will continue; no imminent solution is likely (though see Hey [2006] on perceived progress). But we can learn from this debate and successfully continue both basic and applied research in biodiversity.

Within ornithology, the political and public landscapes still seem to be dominated by the BSC (e.g., del Hoyo et al. 1992–2005, AOU 1998, Dickinson 2003). However, this apparent stability in species concepts does not equal a rigidity in recognized species. Since the widespread adoption of the biological species concept, the number of recognized bird species has grown more rapidly than new species have been described (on the latter, see Banks 2004). Mayr (1946) estimated that there were 8,616 species of birds. Decades later, he raised that estimate to about 9,000 (Mayr 1982). Sibley and Monroe (1990) considered that there were 9,672 bird species, and Dickinson (2003) gave 9,721. Implementation of the BSC resulted in over-lumping (see Peters et al. 1934–1986), and as our understanding of species (and subspecies) limits has improved and more data have become available, many taxa treated as subspecies under the BSC have undergone taxonomic revision and are now recognized as species (compare Dickinson 2003 with Peters et al. 1934–1986; see also Haffer 1997). This trend of increasing numbers of recognized species will continue without a change in species concepts (all of the works cited here use the BSC), even if no more bird species new to science are described. A similar phenomenon has occurred in primate taxonomy (Mace 2004). The point is that continued study of diversity under the BSC is not stymied or frozen in time by adherence to that concept, and some of the complaints against it are being recognized and rectified.

As Hey (2001) pointed out, we have been without a consensus on how to define species over the entire history of evolutionary biology. Ongoing development of this field has probably occurred because our science has progressed toward a working solution that separates the semantic from the empirical aspects of understanding species, an approach explicitly recommended for continued progress (Hey et al. 2003). As major works in evolutionary biology continue to purposely use the BSC (e.g., Avise 2000a, West-Eberhard 2003, Coyne and Orr 2004), basic science is continuing without undue angst over this lack of consensus on what constitutes a species.

## Acknowledgments

We thank G. Spellman for insightful discussions and participation in the early phases of this manuscript and C. Cicero, M. Patten, and J. V. Remsen for helpful comments.

## Literature Cited

Agapow, P.-M., O. R. P. Bininda-Emonds, K. A. Crandall, J. L. Gittleman, G. M. Mace, J. C. Marshall, and A. Purvis. 2004. The impact of species concept on biodiversity studies. Quarterly Review of Biology 79:161–179.

American Ornithologists' Union (AOU). 1957. Check-list of North American Birds, 5th ed. American Ornithologists' Union, Baltimore, Maryland.

American Ornithologists' Union (AOU). 1998. Check-list of North American Birds, 7th ed. American Ornithologists' Union, Washington, D.C.

Arcese, P., M. K. Sogge, A. B. Marr, and M. A. Patten. 2002. Song Sparrow (*Melospiza melodia*). *In* The Birds of North America, no. 704 (A. Poole and F. Gill, Eds.). Birds of North America, Philadelphia.

Avise, J. C. 2000a. Phylogeography: The History and Formation of Species. Harvard University Press, Cambridge, Massachusetts.

Avise, J. C. 2000b. Cladists in wonderland. Evolution 54:1828–1832.

Avise, J. C. 2004. Molecular Markers, Natural History, and Evolution, 2nd ed. Sinauer Associates, Sunderland, Massachusetts.

Avise, J. C., and R. M. Ball, Jr. 1990. Principles of genealogical concordance in species concepts and biological taxonomy. Oxford Surveys in Evolutionary Biology 7:45–67.

Avise, J. C., and D. Walker. 1999. Species realities and numbers in sexual vertebrates: Perspectives from an asexually transmitted genome. Proceedings of the National Academy of Sciences USA 96:992–995.

Avise, J. C., and K. Wollenberg. 1997. Phylogenetics and the origin of species. Proceedings of the National Academy of Sciences USA 94: 7748–7755.

Banks, R. C. 2004. Ornithological nomenclature. Pages 13–25 *in* Handbook of the Birds of the World, vol. 9 (J. del Hoyo, A. Elliott, and D. A. Christie, Eds.). Lynx Edicions, Barcelona, Spain.

Barrowclough, G. F., and N. R. Flesness. 1996. Species, subspecies, and races: The problem of units of management in conservation. Pages 247–254 *in* Wild Mammals in Captivity: Principles and Techniques (D. G. Kleiman, M. E. Allen, K. V. Thompson, S. Lumpkin, and H. Harris, Eds.). University of Chicago Press, Chicago, Illinois.

Barrowclough, G. F., J. G. Groth, L. A. Mertz, and R. J. Guttiérrez. 2005. Genetic structure, introgression, and a narrow hybrid zone between northern and California spotted owls (*Strix occidentalis*). Molecular Ecology 14:1109–1120.

Baum, D. A., and K. L. Shaw. 1995. Genealogical perspectives on the speciation problem. Pages 289–303 *in* Experimental and Molecular Approaches to Plant Biosystematics (P. C. Hoch and A. G. Stephenson, Eds.). Monographs in Systematic Botany, Missouri Botanical Garden, St. Louis, Missouri.

BirdLife International. 2000. Threatened Birds of the World. Lynx Edicions and BirdLife International, Barcelona, Spain, and Cambridge, United Kingdom.

Bock, W. J. 1992. The species concept in theory and practice. Zoological Science 9:697–712.

Brooks, T. M., R. A. Mittermeier, G. A. B. Da Fonseca, J. Gerlach, M. Hoffmann, J. F. Lamoreux, C. G. Mittermeier, J. D. Pilgrim, and A. S. L. Rodrigues. 2006. Global biodiversity priorities. Science 313:58–61.

Bulgin, N. L., H. L. Gibbs, P. Vickery, and A. J. Baker. 2003. Ancestral polymorphisms in genetic markers obscure detection of evolutionarily distinct populations in the endangered Florida Grasshopper Sparrow (*Ammodramus savannarum floridanus*). Molecular Ecology 12: 831–844.

Collar, N. J. 1997. Taxonomy and conservation: Chicken and egg. Bulletin of the British Ornithologists' Club 117:122–136.

Committee on the Status of Endangered Wildlife in Canada (COSEWIC). 2005. Guidelines for recognizing designatable units below the species level (Appendix F5 in the COSEWIC O&P Manual). Committee on the Status of Endangered Wildlife in Canada, Ottawa, Ontario. [Online.] Available at www.dfo-mpo.gc.ca/csas/Csas/Schedule-Horraire/Details/2005/11_Nov/COSEWIC_DU_guidelines_EN.pdf.

Coyne, J. A., and H. A. Orr. 2004. Speciation. Sinauer Associates, Sunderland, Massachusetts.

Cracraft, J. 1983. Species concept and speciation analysis. Pages 159–187 *in* Current Ornithology, vol. 1 (R. F. Johnston, Ed.). Plenum Press, New York.

Cracraft, J. 1989. Speciation and its ontology: The empirical consequences of alternative species concepts for understanding patterns and processes of differentiation. Pages 28–59 *in* Speciation and Its Consequences (D. Otte and J. A. Endler, Eds.). Sinauer Associates, Sunderland, Massachussetts.

Cracraft, J. 1997. Species concepts in systematics and conservation biology—An ornithological viewpoint. Pages 325–339 *in* Species: The Units of Biodiversity (M. F. Claridge, H. A. Dawah, and M. R. Wilson, Eds.). Chapman and Hall, London.

Cracraft, J. 2000. Species concepts in theoretical and applied biology: A systematic debate with consequences. Pages 3–14 *in* Species Concepts and Phylogenetic Theory: A Debate (Q. D. Wheeler and R. Meier, Eds.). Columbia University Press, New York.

Crandall, K. A., O. R. P. Bininda-Emonds, G. M. Mace, and R. K. Wayne. 2000. Considering evolutionary processes in conservation biology. Trends in Ecology and Evolution 15: 290–295.

Cronin, M. A. 1993. Mitochondrial DNA in wildlife taxonomy and conservation biology: Cautionary notes. Wildlife Society Bulletin 21: 339–348.

Cutright, P. R., and M. J. Brodhead. 1981. Elliott Coues: Naturalist and Frontier Historian. University of Illinois Press, Urbana.

Darwin, C. 1859. On the Origin of Species by Means of Natural Selection. John Murray, London.

del Hoyo, J., A. Elliott, J. Sargatal, and D. A. Christie, Eds. 1992–2005. Handbook of the Birds of the World, vols. 1–10. Lynx Edicions, Barcelona, Spain.

de Quieroz, K. 1998. The general lineage concept of species, species criteria, and the process of speciation: A conceptual unification and terminological recommendations. Pages 57–75 *in* Endless Forms: Species and Speciation (D. J. Howard and S. H. Berlocher, Eds.). Oxford University Press, Oxford, United Kingdom.

Dickinson, E. C., Ed. 2003. The Howard and Moore Complete Checklist of the Birds of the World, 3rd ed. Princeton University Press, Princeton, New Jersey.

Endangered Species Act (ESA). 1973. Public Law 93-205, Approved Dec. 28, 1973, 87 Stat. 884, As Amended Through Public Law 107-136, Jan. 24, 2002.

Endler, J. A. 1989. Conceptual and other problems in speciation. Pages 625–648 *in* Speciation and Its Consequences (D. Otte and J. A. Endler, Eds.). Sinauer Associates, Sunderland, Massachussetts.

Funk, D. J., and K. E. Omland. 2003. Species-level paraphyly and polyphyly: Frequency, causes, and consequences, with insights from animal mitochondrial DNA. Annual Review of Ecology, Evolution, and Systematics 34:397–423.

Gibson, D. D., and B. Kessel. 1997. Inventory of the species and subspecies of Alaska birds. Western Birds 28:45–95.

Godfray, H. C. J. 2002. Challenges for taxonomy. Nature 417:17–19.

Grant, P. R., and B. R. Grant. 1992. Hybridization in bird species. Science 256:193–197.

Gray, A. P. 1958. Bird Hybrids: A Check-list with Bibliography. Commonwealth Agricultural Bureaux, Farnham Royal, England.

Green, D. M. 2005. Designatable units for status assessment of endangered species. Conservation Biology 19:1813–1820.

Haffer, J. 1997. Species concepts and species limits in ornithology. Pages 11–24 *in* Handbook of the Birds of the World, vol. 4 (J. del Hoyo, A. Elliott, and J. Sargatal, Eds.). Lynx Edicions, Barcelona, Spain.

Harrison, R. G. 1998. Linking evolutionary pattern and process: The relevance of species concepts for the study of speciation. Pages 19–31 *in* Endless Forms: Species and Speciation (D. J. Howard, and S. H. Berlocher, Eds.). Oxford University Press, Oxford, United Kingdom.

Hebert, P. D. N., M. Y. Stoekle, T. S. Zemlak, and C. M. Francis. 2004. Identification of birds through DNA barcodes. PLoS Biology 2: 1657–1663.

Helbig, A. J., A. G. Knox, D. T. Parkin, G. Sangster, and M. Collinson. 2002. Guidelines for assigning species rank. Ibis 144:518–525.

Hey, J. 2001. The mind of the species problem. Trends in Ecology and Evolution 16:326–329.

Hey, J. 2006. On the failure of modern species concepts. Trends in Ecology and Evolution 21: 447–450.

Hey, J., R. S. Waples, M. L. Arnold, R. K. Butlin, and R. G. Harrison. 2003. Understanding and confronting species uncertainty in biology and conservation. Trends in Ecology and Evolution 18:597–603.

Johnson, N. K., and C. Cicero. 2004. New mitochondrial DNA data affirm the importance of Pleistocene speciation in North American birds. Evolution 58:1122–1130.

Johnson, N. K., J. V. Remsen, Jr., and C. Cicero. 1999. Resolution of the debate over species concepts in ornithology: A new comprehensive biologic species concept. Pages 1470–1482 *in* Acta XXII Congressus Internationalis Ornithologici (N. J. Adams and R. H. Slotow, Eds.). BirdLife South Africa, Johannesburg.

King, M. 1993. Species Evolution: The Role of Chromosome Change. Cambridge University Press, New York.

König, C., F. Weick, and J.-H. Becking. 1999. Owls: A Guide to the Owls of the World. Yale University Press, New Haven, Connecticut.

Levin, D. A. 1979. The nature of plant species. Science 204:381–384.

Mace, G. M. 2004. The role of taxonomy in species conservation. Proceedings of the Royal Society of London, Series B 359:711–719.

Mace, G. M., and N. J. Collar. 2002. Priority-setting in species conservation. Pages 61–73 *in* Conserving Bird Biodiversity: General Principles and Their Application (K. Norris and D. J. Pain, Eds.). Cambridge University Press, London.

Mallet, J. 1995. A species definition for the modern synthesis. Trends in Ecology and Evolution 10:294–299.

Mayden, R. L. 1997. A hierarchy of species concepts: The denouement in the saga of the species problem. Pages 381–424 *in* Species: The Units of Biodiversity (M. F. Claridge, H. A. Dawah, and M. R. Wilson, Eds.). Chapman and Hall, New York.

Mayr, E. 1942. Systematics and the Origin of Species from the Viewpoint of a Zoologist. Columbia University Press, New York.

Mayr, E. 1946. The number of species of birds. Auk 63:64–69.

Mayr, E. 1963. Animal Species and Evolution. Belknap Press of Harvard University Press, Cambridge, Massachusetts.

Mayr, E. 1969. Principles of Systematic Zoology. McGraw-Hill, New York.

Mayr, E. 1982. The Growth of Biological Thought: Diversity, Evolution, and Inheritance. Belknap Press of Harvard University Press, Cambridge, Massachusetts.

Mayr, E. 1993. Fifty years of research on species and speciation. Proceedings of the California Academy of Science 48:131–140.

Mayr, E. 1996. What is a species and what is not? Philosophy of Science 63:262–277.

McKitrick, M. C., and R. M. Zink. 1988. Species concepts in ornithology. Condor 90:1–14.

Moore, W. S., J. H. Graham, and J. T. Price. 1991. Mitochondrial DNA variation in the Northern

Flicker (*Colaptes auratus*). Molecular Biology of Evolution 8:327–344.

Moritz, C. 1994. Defining "evolutionarily significant units" for conservation. Trends in Ecology and Evolution 9:373–375.

Moritz, C. 1996. Use of molecular phylogenies for conservation. Pages 203–216 *in* New Uses for New Phylogenies (P. H. Harvey, A. J. Leigh Brown, J. Maynard Smith, and S. Nee, Eds.). Oxford University Press, Oxford, United Kingdom.

Moritz, C. 2002. Strategies to protect biological diversity and the evolutionary processes that sustain it. Systematic Biology 51:238–254.

Moritz, C., and C. Cicero. 2004. DNA barcoding: Promises and pitfalls. PLoS Biology 2: 1529–1531.

Mumme, R. L., M. L. Galatowitsch, P. G. Jabłoński, T. M. Stawarczyk, and J. P. Cygan. 2006. Evolutionary significance of geographic variation in a plumage-based foraging adaptation: An experimental test in the Slate-throated Redstart (*Myioborus miniatus*). Evolution 60:1086–1097.

Nelson, G. J., and N. I. Platnick. 1980. Multiple branching in cladograms: Two interpretations. Systematic Zoology 29:86–91.

Nixon, K. C., and Q. D. Wheeler. 1990. An amplification of the phylogenetic species concept. Cladistics 6:211–223.

Norris, K., and D. J. Pain, Eds. 2002. Conserving Bird Biodiversity: General Principles and Their Application. Cambridge University Press, Cambridge, United Kingdom.

O'Brien, S. J., and E. Mayr. 1991. Bureaucratic mischief: Recognizing endangered species and subspecies. Science 1187–1188.

Orr, H. A. 2001. The genetics of species differences. Trends in Ecology and Evolution 16: 343–350.

Paetkau, D. 1999. Using genetics to identify conservation units: A critique of current methods. Conservation Biology 13:1507–1509.

Paterson, H. E. H. 1985. The recognition concept of species. Pages 21–29 *in* Species and Speciation (E. S. Vrba, Ed.). Transvaal Museum Monographs, no. 4. Pretoria, South Africa.

Patten, M. A., and P. Unitt. 2002. Diagnosability versus mean differences of Sage Sparrow subspecies. Auk 119:26–35.

Patton, J. L., and M. F. Smith. 1989. Population structure and the genetic and morphological divergence among pocket gopher species (genus *Thomomys*). Pages 284–304 *in* Speciation and Its Consequences (D. Otte and J. A. Endler, Eds.). Sinauer Associates, Sunderland, Massachussetts.

Peters, J. L., et al. 1934–1986. Check-list of Birds of the World, vols. I–XVI. Museum of Comparative Zoology, Harvard University, Cambridge, Massachusetts.

Peterson, A. T. 1998. New species and new species limits in birds. Auk 115:555–558.

Pruett, C. L. 2002. Phylogeography and population genetic structure of Beringian landbirds. Ph.D. dissertation, University of Alaska, Fairbanks.

Pruett, C. L., D. D. Gibson, and K. Winker. 2004. Amak Island Song Sparrows (*Melospiza melodia amaka*) are not evolutionarily significant. Ornithological Science 3:133–138.

Pruett, C. L., and K. Winker. 2005. Northwestern Song Sparrow populations show genetic effects of sequential colonization. Molecular Ecology 14:1421–1434.

Remsen, J. V., Jr. 2005. Pattern, process, and rigor meet classification. Auk 122:403–413.

Roosevelt, T. 1897. The discrimination of species and subspecies. Science 5:879–880.

Ryder, O. A. 1986. Species conservation and systematics: The dilemma of subspecies. Trends in Ecology and Evolution 1:9–10.

Sibley, C. G., and B. L. Monroe, Jr. 1990. Distribution and Taxonomy of Birds of the World. Yale University Press, New Haven, Connecticut.

Sites, J. W., Jr., and J. C. Marshall. 2004. Operational criteria for delimiting species. Annual Review of Ecology, Evolution, and Systematics 35:199–227.

Talbot, S. L., and G. F. Shields. 1996. Phylogeography of brown bears (*Ursus arctos*) of Alaska and paraphyly within the Ursidae. Molecular Phylogenetics and Evolution 5: 477–494.

Taylor, H. L., J. M. Walker, J. E. Cordes, and G. J. Manning. 2005. Application of the evolutionary species concept to parthenogenetic entities: Comparison of postformational divergence in two clones of *Aspidoscelis tesselata* and between *Aspidoscelis cozumela* and *Aspidoscelis maslini* (Squamata: Teiidae). Journal of Herpetology 39:266–277.

Templeton, A. R. 1989. The meaning of species and speciation: A genetic perspective. Pages 3–27 *in* Speciation and Its Consequences (D. Otte and J. A. Endler, Eds.). Sinauer Associates, Sutherland, Massachusetts.

Villaseñor, J. L., G. Ibarra-Manríquez, J. A. Meave, and E. Ortíz. 2005. Higher taxa as surrogates of plant biodiversity in a megadiverse country. Conservation Biology 19:232–238.

West-Eberhard, M. J. 2003. Developmental Plasticity and Evolution. Oxford University Press, Oxford, United Kingdom.

Wheeler, Q. D., and R. Meier, Eds. 2000. Species Concepts and Phylogenetic Theory: A Debate. Columbia University Press, New York.

Wiley, E. O. 1978. The evolutionary species concept reconsidered. Systematic Biology 27:17–26.

Williams, P. H., K. J. Gaston, and C. J. Humphries. 1997. Mapping biodiversity value worldwide: Combining higher-taxon richness from different groups. Proceedings of the Royal Society of London, Series B 264:141–148.

Zink, R. M. 1996. Species concepts, speciation, and sexual selection. Journal of Avian Biology 27: 1–6.

Zink, R. M. 2004. The role of subspecies in obscuring avian biological diversity and misleading conservation. Proceedings of the Royal Society of London, Series B 271:561–564.

Zink, R. M., and M. C. McKitrick. 1995. The debate over species concepts and its implications for ornithology. Auk 112:701–719.

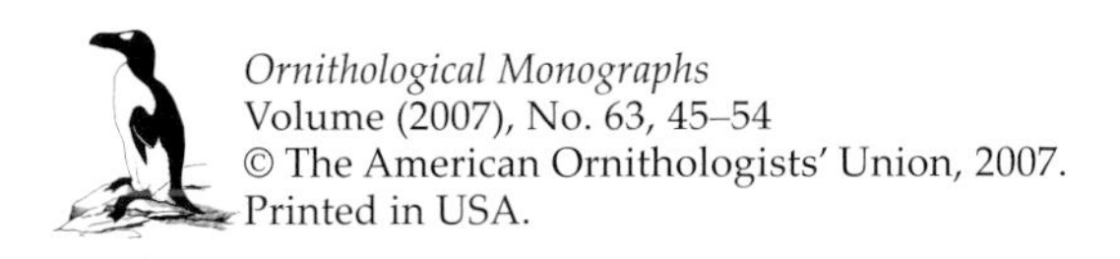

*Ornithological Monographs*
Volume (2007), No. 63, 45–54

Printed in USA.

CHAPTER 4

# NAMED SUBSPECIES AND THEIR SIGNIFICANCE IN CONTEMPORARY ORNITHOLOGY

James D. Rising[1]

*Department of Zoology, University of Toronto, Toronto, Ontario M5S 3G5, Canada*

Abstract.—Subspecies, or geographic races, are diagnosable populations that, at least during the breeding season, are largely allopatric with other subspecies of the same species. In attempts to give objectivity to the subspecies concept, arbitrary rules have been applied for the recognition of subspecies (e.g., the "75% rule," whereby 75% of the individuals should be identifiable to subspecies; there are several other rules). As a case study, I examined the usefulness of the subspecies concept in describing geographic variation of a polytypic American songbird, the Savannah Sparrow (*Passerculus sandwichensis*). About 21 subspecies of this species have been recognized in the taxonomic literature, but much of the geographic variation is clinal. I argue that there is little value in subdividing a clinal continuum into different subspecies. Rather, the use of subspecies is best restricted to distinctive, and usually geographically isolated, populations. I show that this has been done for only a few of the named subspecies of Savannah Sparrows. *Received 31 July 2006, accepted 6 March 2007.*

Resumen.—Las subespecies, o razas geográficas, son poblaciones generalmente alopátricas (al menos durante la época de reproducción) y que se pueden diferenciar claramente de otras subespecies de la misma especie. Se han propuesto diferentes reglas para asignar individuos a una determinada subespecie, e.g. "la regla del 75%" (75% de los individuos tienen que ser identificables como subespecie), en un intento de dar objetividad al concepto de subespecie. Como caso de estudio, discuto la utilidad del concepto de subespecie para describir la variación geográfica de *Passerculus sandwichensis*. Se han reconocido cerca de 21 subespecies, pero gran parte de la variación geográfica es clinal. Argumento que es de poco valor el subdividir una clina continua en subespecies diferentes. Se debe restringir el uso de subespecies para poblaciones bien diferenciadas y aisladas geográficamente. Muestro que ésto se ha hecho sólo para unas pocas de las subespecies de *Passerculus sandwichensis*.

In recent years, many biologists have used intraspecific geographic variation to test hypotheses about adaptation and evolution, and named subspecies have reflected this variation. For example, Møller and Cuervo (1998) compared feather ornamentation in birds to test the hypothesis that sexual selection promotes speciation and found that ornamented species had more subspecies than non-ornamented species—which suggests an association between subspeciation and ornamentation. Likewise, Sol et al. (2005) examined brain size relative to body size in Holarctic passerines, to test the hypothesis that behavioral changes are an important driver of evolutionary diversification, and found that species with large relative brain size have undergone more extensive subspecific diversification. It is clearly important for these studies that named subspecies more or less accurately reflect units of intraspecific diversification.

Wilson and Brown (1953) identified several problems with the subspecies concept as then applied. Among these was the arbitrary lower limit of the "distinctiveness" of subspecies (i.e., how distinctive must a population be to earn a subspecific trinominal name?). There are several arbitrary rules, the most common being the "75% rule" (Amadon 1949), though efforts have been made to apply more rigorous diagnosability rules to the classification of subspecies (e.g., Patten and Unitt 2002, Remsen 2005, Cicero and

[1]E-mail: rising@zoo.utoronto.ca

Johnson 2006). In 1982, John Wiens, then the editor of *The Auk*, solicited comments from several American avian systematists about the utility and contemporary application of the subspecies concept. Wiens (1982:593) wrote:

> My charge to them was framed as a series of questions: Is the concept just a tool of classification that is no longer of much use? Can or should the concept be revised to make it more compatible with contemporary views in population biology? Do subspecies exist, as real biological units?

Among the contributors to this forum were George Barrowclough, Frank Gill, Ned Johnson, Ernst Mayr, Burt Monroe, Robert Storer, and Richard Zusi. Some 25 years later, these questions are still debated and deserve revisiting.

Ned Johnson (1982:605) identified four "distinctly positive elements" about subspecies:

> [1] [S]ubspecies tell us about the migratory routes and wintering areas of populations of birds that represent distinct portions of the breeding range of that species.
>
> [2] Some subspecies also provide indisputable evidence for the early stages of allopatric speciation in relation to environmental barriers.
>
> [3] Subspecies names alert researchers of whatever stripe to geographic forms with potential differences in features additional to those by which they were initially characterized.
>
> [4] Systematists...can profit by looking closely at subspecies named on the basis of particular kinds of characters. After all, some of these "subspecies" will turn out after careful study to be full species.

Most of the ornithologists who contributed to the forum concurred with Monroe (1982:608) when he wrote:

> It seems an abuse of the naming process to create a name for a population that may differ in one slight character (and even then, subspecies may be named where only about two-thirds of the individuals can be distinguished, based on the most liberal interpretation of the "75% rule").

He went on:

> [S]ubspecific names should *not* be used to describe populations differing only through smooth clines reflecting general primary intergradation.... Subspecies...should be used in two situations: (1) allopatric populations where definition of the populations is clear, distinct, and total (or very nearly so); and (2) situations where secondary contact between distinct populations has occurred and the zone of intergradation is relatively narrow.

Barrowclough (1982:602) wrote:

> ...I see some use in the formal naming of subspecies, but only if standards become much more rigorous. First, we need to acknowledge that a useful subspecies concept will have as a goal the same objective as other taxonomic categories—*predictiveness*.

If a population is distinctive in coloration, size, or some other character, is it also distinctive in other ways? Thus, subspecies are most useful if they represent distinct gene pools that predict variation in traits not originally considered.

## Subspecies versus Species

Cracraft (1983, 1997) and others have suggested that "phylogenetic species," based on cladistic analyses that ignore interbreeding among populations, should replace "biological species" defined on the basis of observed or presumed reproductive isolation—and that subspecies, generally recognized within biological species, are not useful taxonomic entities (e.g., McKitrick and Zink 1988, Zink 2004). Others (e.g., Avise and Ball 1990, Johnson et al. 1999, Remsen 2005, Winker et al. 2007) have argued against this because it ignores the importance of gene flow, which runs counter to theories of speciation and population differentiation. If we embrace biological species, some taxa at the species and subspecies levels will be paraphyletic (e.g., Funk and Omland 2003), which most advocates of phylogenetic species find unacceptable. Although the boundary between subspecies and species can be fuzzy, the advent of molecular tools has led some to elevate subspecies to species on the basis of arbitrary levels of genetic distinctiveness (e.g., Hebert et al. 2004, Kerr et al. 2007). These approaches are often based on single-locus, selectively neutral markers, which reflect gene trees but not necessarily species trees (Johnson and Cicero 2004, Moritz and Cicero 2004, Winker et al. 2007). Concordance between phenotypic data (the basis for most named subspecies) and genotypic data, combined with other indicators of reproductive isolation (e.g., behavior), serve as indicators of gene flow and thus of species status between populations (Johnson et al. 1999).

## Morphological and Molecular Diagnosability of Subspecies

Patten and Unitt (2002) and Cicero and Johnson (2006) emphasized the need for objective criteria and increased rigor in delimiting subspecies morphologically. Furthermore, Cicero and Johnson (2006), among others, emphasized the importance of using only breeding birds in describing geographic variation. With migratory birds, this raises a significant technical problem because often the breeding plumage is worn, thus obscuring or changing their appearance. Traditionally, taxonomists have preferred to describe species and subspecies in "fresh" plumage, although in practice few descriptions have been based on such material. In many cases, most individuals molt before leaving their breeding grounds, so there is a narrow window of time when specimens in fresh plumage can be collected from their breeding grounds. Furthermore, molt sequences of many species are poorly known, and at least some molting may take place during migration or on the wintering grounds (Rohwer and Johnson 1992, Rising and Beadle 1996, Pyle 1997). Also, gonadal changes precede molt, so it is not possible to know whether a recently molted specimen represents a local breeder or a migrant. Lastly, if the subspecies' names can be applied only to freshly molted individuals, their utility is greatly reduced. For example, if the plumage of individuals that have completed their migration is worn, how can it be matched with breeding material to clarify migration routes?

Zink (2004) argued that subspecies (or phylogenetic species) should be the units to be considered when making conservation decisions, which would be wise if named subspecies represent distinctive gene pools. However, in a mitochrondrial DNA (mtDNA) analysis of subspecies, Zink (2004) found that only ~3% of the named subspecies of birds coincided with distinct, monophyletic gene lineages. He based this result on a meta-analysis of 41 species, with a distinct bias toward Nearctic and Palearctic taxa (220 out of 230 of the subspecies examined). Because mtDNA loci within lineages are believed to evolve mostly by drift and float among populations through matrilineal inheritance, mtDNA loci give, at best, limited insight into phylogenetic and phylogeographic history (Edwards and Beerli 2000, Arbogast et al. 2002, Funk and Omland 2003). Furthermore, measures of mtDNA diversity likely do not reflect quantitative genetic variation for adaptive traits (Reed and Frankham 2001). Therefore, there is no *a priori* reason to expect mtDNA gene trees to precisely reflect either population histories or population boundaries delimited by genes that affect color, size, or shape—the traits upon which most named subspecies have been described. Despite these caveats, Phillimore and Owens (2006) repeated Zink's analysis using taxa from a broader global region and found that 36% of the subspecies represented distinct phylogenetic lineages. Not surprisingly, more island-dwelling subspecies than subspecies with a continental distribution were monophyletic. This is consistent with what I postulate about allopatric speciation: in patchy environments, populations are often smaller and gene flow and its blending effects among patches reduced, allowing local differentiation to evolve more quickly.

## The Savannah Sparrow as a Case Study

Here, I assess both the historical application and the usefulness of the subspecies concept by using the Savannah Sparrow (*Passerculus sandwichensis*) as a case study. The Savannah Sparrow is a widespread North American species that has been divided into at least 17 to 21 subspecies (American Ornithologists' Union [AOU] 1957, Paynter 1970), and has been the subject of numerous studies (Wheelwright and Rising 1993). As with many avian species, most currently named subspecies were described in the first half of the 20th century, at a time when rigorous statistical analyses (e.g., *t*-tests and analyses of variance [ANOVA]) were just being developed. Specifically, my objective is to illustrate many of the problems with current subspecific taxonomy, while addressing what to do with these named subspecies: (1) Should we accept them as named? (2) Should we revisit these with an eye to making their descriptions more consistent and rigorous?

Or (3) should we do away with intraspecific groups and nomenclature?

Peters and Griscom (1938) is the only monographic study of the geographic variation and intraspecific taxonomy of the Savannah Sparrow that deals with specimens taken from most parts of the species' range. Their treatment of Mexican populations, however, was based on few specimens (and still fewer breeding specimens) and therefore was necessarily preliminary. Later, Aldrich (1940) discussed variation among eastern North American Savannah Sparrows; van Rossem (1947) published a study of geographic variation of the saltmarsh-dwelling coastal Belding's and Large-billed sparrows of southern California, Baja California, Sonora, and Sinaloa; and Hubbard (1974) analyzed variation among the Savannah Sparrows of the southwestern United States, inland Mexico, and Guatemala. More recently, I published information on geographic variation in size and shape of Savannah Sparrows (Rising 2001), and Zink et al. (1991, 2005) reported on molecular differences among populations of Savannah Sparrows. Finally, I have quantified variation among populations of Savannah Sparrows on the basis of pattern and coloration of plumage (J. D. Rising unpubl. data). All these studies contain quantitative data that support the statements made here.

Peters and Griscom (1938) based their subspecies on qualitative descriptions of size, bill shape, and morphology. Thus, *P. s. labradorius* was diagnosed as "a dark Savannah Sparrow with relatively stout bill; its depth more than half the length of the culmen" (p. 452), and *P. s. sandwichensis* as the "largest (average) of the races… with a long and proportionately slender bill" (p. 448). Even new subspecies described in that monograph were not characterized in a quantitative way. For example, *P. s. oblitus* was described as "a medium sized gray Savannah Sparrow with relatively stout bill, its depth more than half the length of the culmen" (p. 455). It was "similar to *P. s. savanna*…but grayer throughout," (p. 455) and it "intergrades with [*P. s.*] *nevadensis* where the forms meet" (p. 457). Likewise, van Rossem (1947) gave only qualitative descriptions of the saltmarsh Savannah Sparrows of Mexico, even when describing a new subspecies, *P. s. magdalenae*. He wrote (van Rossem 1947:103):

> This race is the culmination of the strongly yellow-browed, peninsular Savannah Sparrows with relatively slender bills which average less…than 7.0 millimeters in depth at base. It forms a good connecting link between the smaller-billed, more northern *guttatus* and the larger-billed *rostratus* group of the continental mainland and the San Benito Islands….

I found that differences in bill proportions among populations of non-saltmarsh ("typical") Savannah Sparrows were slight and that bill size varied clinally (Rising 2001); my analyses were based on large samples (generally >40 birds from each) of breeding birds from 55 localities from virtually throughout the species' range. From these analyses, it is unlikely that any discrete populations of Savannah Sparrows exist solely on the basis of bill size or shape.

Although some of the 21 or so named subspecies of Savannah Sparrows have been described or at least characterized by average differences—a criterion that Patten and Unitt (2002) argued is insufficient for subspecies diagnosability—most, like *P. s. magdalenae* from Bahía Magdalena and *P. s. wetmorei* from Hacienda Chancol, Guatemala, have been described on the basis of impressions of differences that were not quantified in any way. Van Rossem (1938) described *P. s. wetmorei* from only five specimens collected in June 1897, and there seem to be no breeding specimens subsequently taken from south of Mexico. Hubbard (1974:14) noted that "Guatemalan specimens differ from Southwest specimens in being darker and ruddier brown above, with the streaking more extensive and darker; the yellow of the superciliary also tends to be darker and more extensive." In his description of *P. s. magdalenae*, van Rossem (1947:102) measured 16 males and 4 females and characterized this subspecies as similar to *P. s. guttatus* from Laguna San Ignacio "but lighter and more greenish (less grayish) olive; dorsal markings more prominent…due to lighter edgings." In his color plate (van Rossem 1947), *P. s. magdalenae* and *P. s. guttatus* appear to be very similar, as they do in the field (J. D. Rising pers. obs.).

None of the subspecies of Savannah Sparrow have been named following the mandates of any rule, and indeed, some have not been formally described (although perhaps described long after being first named). For example, Latham (AOU 1957) described the Sandwich Bunting from the Aleutian Islands, later named *Emberiza sandwichensis* by Gmelin in 1789; and in 1811, Wilson described *Fringilla savanna* from Savannah, Georgia. When these were synonymized, they became different subspecies of the same species, even though there was no formal description of their differences. In many instances, the subspecies represent only points on a clinal continuum (Rising 2001, Rising et al. unpubl. data). Only Hubbard (1974) attempted to delimit subspecies of Savannah Sparrows on the basis of the separability of adjacent populations. Several subspecies (e.g., *P. s. princeps, P. s. bradburyi, P. s. guttatus,* and *P. s. rostrata*) have been based on material taken from migrating or wintering individuals, and others have been based, in whole or in part, on nonbreeding individuals. Subspecific names should be given to populations or groups of populations that occupy a distinct breeding range and that are diagnosably distinct from other such populations (Mayr and Ashlock 1991, Patten and Unitt 2002, Cicero and Johnson 2006).

Among non-saltmarsh sparrows, the large and pallid Savannah Sparrows from Sable Island ("Ipswich Sparrow," *P. s. princeps;* Rising 2001) are the only ones that are consistently separable from all others. These birds are essentially 100% separable from mainland Savannah Sparrows using either size or color, although their mtDNA haplotypes are not distinctive (Zink et al. 2005). Among saltmarsh Savannah Sparrows, the birds from coastal Sinaloa and Sonora (*rostratus* group) are large and large-billed (Rising 2001) and differ in coloration (Rising et al. unpubl. data) from those along the coast of southern California and Baja California (*P. s. beldingi*); in my analyses, there is virtually no overlap between these sets of populations. Size and color variation along the coast of Sinaloa and Sonora is clinal, although color differences completely separate birds from Puerto Peñasco (*P. s. rostratus*) and Bahía Kino (*P. s. atratus*); birds from the Cabo Lobos region of Sonora are said to be intermediate (van Rossem 1947). Similarly, variation along the Pacific Coast is clinal in both color and size, although there is a gap between Morro Bay and Humboldt Bay (*P. s. alaudinus*) in California, and another between Guerrero Negro (*P. s. anulus*) and Bahía Magdalena (*P. s. magdalenae*) on the Baja California Peninsula. Lastly, Savannah Sparrows from Isla San Benito (*P. s. sanctorum*) are distinct from mainland birds morphologically (van Rossem 1947, J. D. Rising pers. obs.), in coloration (Rising et al. unpubl. data), and behaviorally (specimens collected in late April 1999 showed birds on different breeding cycles: laying eggs at Guerrero Negro, and clearly prereproductive on Isla San Benito). On the basis of size and coloration, van Rossem (1947) considered this population to be allied with those of the west coast of Sonora and Sinaloa. However, mtDNA analyses (Zink et al. 2005) do not support this conclusion, but rather ally them with Pacific Coast birds, as would be reasonable on geographical grounds. This suggests that the similarity in bill size between birds from San Benito and those from coastal Sinaloa and Sonora is attributable to convergence.

With regard to the question of what to do with these named subspecies, it is clear that we should neither accept them as named nor eliminate them altogether. Although many of the named subspecies clearly are not diagnosable, there is strong geographic variation that merits formal taxonomic recognition. On the basis of my research (Rising 2001; Zink et al. 1991, 2005; Rising et al. unpubl. data), which has led to a re-evaluation of these subspecies, I recommend that six subspecies be recognized with the following taxonomic synonymies:

### *Passerculus sandwichensis sandwichensis* (Gmelin)

*Emberiza sandwichensis* Gmelin, Syst. Nat., 1, pt. 2, 1789, pl 875. Based on the Sandwich Bunting of Latham, Gen. Syn., vol. 2, pt. 1, p. 202 (In Unalaschca et sinu Sandwich = Unalaska, Alaska).

*Fringilla savanna* Wilson, 1811, Amer. Ornith., 3, p. 55, pl. 22, fig. 3 (Savannah, Georgia).

*Passerculus alaudinus* Bonaparte, 1853, Compt. Rend. Acad. Sci. Paris, 37, p. 918 (Californie = San Francisco).

*Passerculus anthinus* Bonaparte, 1853, Compt. Rend. Acad. Sci. Paris, 37, p. 920 (Kodiak = Kodiak Island, Alaska).

*Passerculus anthinus* (not of Bonaparte, 1853) Baird, 1858, Rep. Pacific R. R. Surv., ix, p. 445 (San Francisco, Benicia, and Petaluma, California).

*Passerculus sandwichensis bryanti* Ridgway, 1885, Proc. U.S. Nat. Mus., 7, p. 157.

*Ammodramus (Passerculus) sandwichensis wilsonianus* Coues, 1897, Auk, 14, p. 93 (new name for *Fringilla savanna* Wilson).

*Ammodramus sandwichensis brunnescens* Butler, 1888, Auk, 5, p. 265 (Valley of Mexico, Mexico).

*Passerculus sandwichensis brunnescens* Oberholser, 1930, Sci. Pub. Cleveland Mus. Nat. Hist., 1, p. 110 (Mexico).
*Passerculus sandwichensis labradorius* Howe, 1901, Contr. North Amer. Orn., vol. 1, Oct. 14, 1901, p. 1 (Lance [L'Anse] au Loup, Labrador).
*Passerculus sandwichensis nevadensis* Grinnell, Univ. California Publs. Zool. 5, no. 9, Feb. 21, 1910, p. 312 (Soldier Meadows, Humboldt County, Nevada).
*Passerculus sandwichensis brooksi* Bishop, 1915, Condor, 17, no. 5, Oct. 10, 1915 (Chilliwack, British Columbia).
*Passerculus sandwichensis bradburyi* Figgins, 1918, Proc. Colorado Mus. Nat. Hist., 2 no. 1, p. 2 (James Island, South Carolina).
*Passerculus sandwichensis campestris* Taverner, 1932, Proc. Biol. Soc. Washington, 45, p. 201 (Red Deer, Alberta).
*Passerculus sandwichensis oblitus* Peters and Griscom, 1938, Bull. Mus. Comp. Zool., 80, no. 13, p. 454 (Fort Churchill, Manitoba).
*Passerculus sandwichensis crassus* Peters and Griscom, 1938, Bull. Mus. Comp. Zool., 80, no. 13, p. 459 (Sitka, Alaska).
*Passerculus sandwichensis wetmorei* van Rossem, 1938, Bull. Brit. Ornith. Club, 58, p. 129 (Hacienda Chancol, 10,000 ft., Guatemala).
*Passerculus sandwichensis mediogriseus* Aldrich, 1940, Ohio J. Sci., 40, p. 4 (Andover, Ashtabula County, Ohio).
*Passerculus sandwichensis rufofuscus* Camras, 1940, Publ. Field Mus. Nat. Hist., Zool. Ser., 24, no. 15, p. 159 (Babicora, Chihuahua).

**Passerculus sandwichensis princeps** Maynard

*Passerculus princeps* Maynard, Amer. Nat., 6, no. 10, Oct. 1872, p. 637 (Ipswich, Massachusetts).

**Passerculus sandwichensis beldingi** Ridgway

*Passerculus beldingi* Ridgway, Proc. U.S. Nat. Mus., 7 (Feb. 25), 1885, p. 516 (San Diego, Cal[ifornia]).
*Passerculus halophilus* (not *Ammodramus halophilus* McGregor) Bancroft, 1927, Condor, 29, p. 56 (Scammon Lagoon, Baja California).
*Passerculus rostratus halophilus* Grinnell, 1928, Univ. Calif. Publs. Zool., 32, p. 163 (Scammon Lagoon, Baja California).
*Passerculus rostratus anulus* Huey, 1930, Trans. San Diego Soc. Nat. Hist., 6, no. 10, p. 204 (Scammon Lagoon, Lower California = Baja California).
*Passerculus sandwichensis anulus* Oberholser, 1930, Sci. Publs. Cleveland Mus. Nat. Hist., 1, p. 110 (Lower California); van Rossem, 1930, Trans. San Diego Soc. Nat. Hist., 6, p. 219 (Scammon Lagoon, Lower California).

**Passerculus sandwichensis sanctorum** Ridgway

*Passerculus sanctorum* Ridgway, Proc. U.S. Nat. Mus., 5, Apr. 3, 1883, p. 538 (Island of San Benito, Pacific coast of Lower California).
*Ammodramus (Passerculus) sanctorum* Coues, 1897, Auk, 14, p. 92 (San Benito Islands, Baja California).
*Passerculus rostratus sanctorum* Ridgway, 1901, Bull. U.S. Nat. Mus., 50, Part 1, p. 200 (San Benitos Islands).
*Passerculus rostratus guttatus* (not of Lawrence) Oberholser, 1919, Ohio J. Sci., 19, p. 349 (San Benito Islands).

**Passerculus sandwichensis guttatus** Lawrence

*Passerculus guttatus* Lawrence, Ann. Lyc. Nat. Hist. New York, 8, 1867, p. 473 (Lower California, San José [del Cabo]).
*Ammodramus halophilus* McGregor, 1898, Auk, 15, p. 265 (Abreojos Point [= Pond Lagoon], Lower California = Baja California Sur).
*Passerculus rostratus halophilus* Ridgway, 1901, Bull. U.S. Nat. Mus., 50, Part 1, p. 202 (Abreojos Point).
*Passerculus sandwichensis halophilus* van Rossem, 1930, Trans. San Diego Soc. Nat. Hist., 6, p. 219 (Abreojos Point south to Magdalena Bay).
*Passerculus sandwichensis magdalenae* van Rossem, 1947, Condor, 49, p. 97 (North Estero, Magdalena Bay, Baja California = Baja California Sur).

**Passerculus sandwichensis rostratus** (Cassin)

*Emberiza rostrata* Cassin, Proc. Acad. Nat. Sci. Philadelphia, Sept.–Oct. (Dec. 31) 1852, p. 184 (sea shore at San Diego, California).
*Ammodramus rostratus* Cassin, 1855, Ill. Birds Calif., Texas, etc., p. 226 (San Diego, Santa Barbara, San Pedro).

*Passerculus rostratus* Grinnell, 1905, Auk, 22, p. 16 (not breeding in California).
*Passerculus sandwichensis rostratus* van Rossem, 1930, Trans. San Diego Soc. Nat. Hist., 6, p. 219 (delta of Colorado River in Lower California and Sonora).
*Passerculus sandwichensis atratus* van Rossem, 1930, Trans. San Diego Soc. Nat. Hist., 6, p. 218 (Tobari Bay, Sonora).

## Conclusions

Are subspecies useful according to the criteria that Johnson (1982) listed? First, can subspecies help us understand patterns of migration? The answer is a qualified "yes," if the subspecies are clearly defined and readily identifiable. In Savannah Sparrows, for example, the large, pale, distinctive birds that breed on Sable Island, Nova Scotia, likely winter along the Atlantic Coast, rarely away from the sand dunes and dune grass. However, many of the other populations, such as the dark and supposedly large *P. s. labradorius* from Labrador, are more difficult to follow. Although dark Savannah Sparrows are common in winter on the Gulf Coast of Texas, which suggests that they come from breeding populations in northeastern Canada, whether they are from Labrador is difficult to determine using conventional morphological criteria. To quote Todd (1963:673), in 1901

> Howe undertook to separate the Labrador birds as *labradorius*. He had just three Labrador specimens, and he based his new race mainly on the supposed larger size of the northern birds. Townsend and Allen (1907), however, relying on Oberholser's positive statement (in litt.) that their Labrador specimens were virtually indistinguishable from southern birds, declined to recognize the race, and this example was followed by most subsequent writers, down to Bangs (1930) who contended that in spite of all adverse criticism, 'the large dark Savannah Sparrow of Labrador is an excellent race.' [Austin (1932)], after examining and comparing his series of freshly collected specimens from the Labrador coast, reached the same conclusion. He defined its characters with considerable precision, stressing its general darker coloration, but discounting its supposed larger size as compared with southern specimens [note that this race was originally defined on the basis of size, and subsequently became defined on the basis of coloration].... Peters and Griscom (1932), with a series of 150 specimens before them, were able to thoroughly establish its validity, *although the majority of their specimens came from south of Labrador* [italics mine]."

Peters and Griscom also described a new subspecies, *P. s. oblitus*, with the type from Churchill, Manitoba. Todd (1963:673) commented,

> ...*labradorius* and *oblitus* are amazingly alike, but are distinguishable in series of seasonally comparable specimens. With only the two type-specimens in hand, I doubt that anyone would have ever considered them to belong to two different races.

On the basis of large series of specimens (Rising 2001, Rising et al. unpubl. data), *labradorius* appears to be darker, on average, than birds from farther to the west, and also slightly larger. However, as noted by Todd (1963), *labradorius* and *oblitus* are not distinguishable with small samples. Although geographic variation exists, the pattern is clinal. Thus, a dark Savannah Sparrow on the Gulf Coast of Texas may well have come from Labrador, but it also may have originated from any place in the James Bay or Hudson Bay lowlands. Mengel (1965) recognized a similar problem with American Robins (*Turdus migratorius*) wintering in Kentucky, where darker individuals could not be traced with certainty to Newfoundland.

Johnson's (1982) other three criteria can be considered together: Do subspecies represent early stages of allopatric speciation? Do they alert researchers to look for other differences? And can some subspecies, with further study, be found to be "good species"? The answer to all these questions, again, is a qualified "yes." If the subspecies have been described rigorously on the basis of sufficient breeding material, they can help us understand evolution (adaptations to local environments) and speciation. Again, to return to Savannah Sparrows, recent studies (Rising 2001, Rising et al. unpubl. data) have shown that some groups of populations are morphologically distinctive: "Ipswich" Sparrows breeding on Sable Island, Nova Scotia; "Belding's" Sparrows along the Pacific Coast of California and Baja California; "Large-billed" Sparrows on the coast of Sonora and Sinaloa; and the "San Benito" Sparrow on the Islas San Benito, Baja California. Genetic data (Zink et al. 1991, 2005) showed that west-coast birds form a distinct clade that contains no "typical" Savannah Sparrows, but that the "Ipswich" Sparrow is a "typical" Savannah Sparrow at the mtDNA level. Although "Ipswich" Sparrows are distinctive morphologically and behaviorally, doubtless those differences are based on genetic variation that has not yet been investigated. Likewise, there is no phylogeographic structure in mtDNA among "Belding's," "Large-billed," and "San Benito" sparrows, although there are morphological differences among them. These morphological "subspecies" have alerted us to look for other differences—which in some cases we have found, and in others not. Are these four groups incipient "biological" species? No doubt. They are mostly or completely allopatric, so the opportunity for interbreeding among populations is minimal. Should we recognize some of these subspecies as different species on the basis of the information that we have at present? I think that one could, but that becomes a matter of taxonomic judgment. The named subspecies, in some cases, have alerted us to the questions.

## The Future

Many subspecies have been described on the basis of few specimens (in some cases, only one), and some on the basis of nonbreeding individuals. Clearly, an arbitrary rule (e.g., 75% or 95% diagnosability) cannot be applied to these without increased samples. Yet many agree that there is some virtue to the subspecies concept if subspecies are based on sufficient data (e.g., Phillimore and Owens 2006, Winker et al. 2007). Those faced with subspecies-level taxonomy need to have a sound basis for revision. Most named subspecies of North American birds are poorly differentiated, and many represent points on a cline. A large number is based on small samples, perhaps of migrants or wintering individuals, and their description is not based on a rigorous analysis of variation. Also, most of these subspecies will not be restudied. Those charged with revising taxonomies should re-evaluate the original descriptions and, when possible, examine material in museums. If a subspecies is defined on the basis of reasonably large samples of breeding birds (e.g., ≥10 specimens from each locality, preferably ≥25; J. D. Rising unpubl. data), combined with at least one diagnosable phenotypic character and preferably two or more, then it should be retained at least until an updated, quantitative study is published. If not, the subspecies name should be stricken from the books. Importantly, subspecies can mislead people about the amount of real geographic variation; for example, geographic variation among populations of Savannah Sparrows is strong but does not conform to 17–21 diagnosable units. In some cases (e.g., "Ipswich" Sparrow), the subspecies are so distinct that they should be retained even if not adequately defined by today's standards. Likewise, subspecies that are diagnosable morphologically but not by mtDNA should be retained because phenotypic differences are more likely than mtDNA to be the result of selection (Mumme et al. 2006). If we restrict subspecies to those that "predict" variation in ways not used in the original description, as Barrowclough (1982) demanded, the subspecies designation is useful. Otherwise, it is of no value and is potentially misleading.

## Acknowledgments

I am greatly indebted to the many people who helped collect samples of the Savannah Sparrows used in my studies of Savannah Sparrows. These include J. M. Torres Ayala, A. Castellanos, F. Martinez, M. Martinez, J. Dick, H. Garcia, R. Harris, A. Jiménez, I. McLaren, P. McLaren, M. Torres-Morales, D. Niles, D. Rising, T. Rising, S. Rohwer, E. Sanchez, P. Schueler, J. Shields, F. Viramontes, and especially H. Medina and F. Schueler. J. Mannone greatly helped me by recording and entering data. C. Cicero, J. V. Remsen, T. Rising, and two anonymous reviewers provided helpful criticism of the manuscript. The cooperation of many national, state, and provincial governments, especially the governments of Canada, Mexico, and the United States, is greatly appreciated. This research has been supported by a grant from the Natural Science and Engineering Research Council of Canada.

## Literature Cited

Aldrich, J. W. 1940. Geographic variation in eastern North American Savannah Sparrows (*Passerculus sandwichensis*). Ohio Journal of Science 40:1–8.

Amadon, D. 1949. The seventy-five per cent rule for subspecies. Condor 51:250–258.

American Ornithologists' Union. 1957. Check-list of North American Birds, 5th ed. American Ornithologists' Union, Baltimore, Maryland.

American Ornithologists' Union. 1998. Check-list of North American Birds, 7th ed. American Ornithologists' Union, Washington, D.C.

Arbogast, B. S., S. V. Edwards, J. Wakely, P. Beerli, and J. B. Slowinski. 2002. Estimating divergence times from molecular data on phylogenetic and population genetic timescales. Annual Review of Ecology and Systematics 33:707–740.

Austin, O. L., Jr. 1932. The Birds of Newfoundland Labrador. Memoirs of the Nuttall Ornithological Club, no. 7.

Avise, J. C., and R. M. Ball, Jr. 1990. Principles of genealogical concordance in species concepts and biological taxonomy. Oxford Surveys in Evolutionary Biology 7:45–67.

Bangs, O. 1930. Types of birds now in the Museum of Comparative Zoology. Bulletin of the Museum of Comparative Zoology 70:147–426.

Barrowclough, G. F. 1982. Geographic variation, predictiveness, and subspecies. Auk 99: 601–603.

Cicero, C., and N. K. Johnson. 2006. Diagnosability of subspecies: Lessons from Sage Sparrows (*Amphispiza belli*) for analysis of geographic variation in birds. Auk 123:266–274.

Cracraft, J. 1983. Species concept and speciation analysis. Pages 159–187 *in* Current Ornithology, vol. 1 (R. F. Johnston, Ed.). Plenum Press, New York.

Cracraft, J. 1997. Species concepts in systematics and conservation biology—An ornithological viewpoint. Pages 325–339 *in* Species: The Units of Biodiversity (M. F. Claridge, H. A. Dawah, and M. R. Wilson, Eds.). Chapman and Hall, London.

Edwards, S. V., and P. Beerli. 2000. Perspective: Gene divergence, population divergence, and the variance in coalescence time in phylogeographic studies. Evolution 54:1839–1854.

Funk, D. J., and K. E. Omland. 2003. Species-level paraphyly and polyphyly: Frequency, causes, and consequences, with insights from animal mitochondrial DNA. Annual Review of Ecology, Evolution, and Systematics 34: 397–423.

Gill, F. B. 1982. Might there be a resurrection of the subspecies? Auk 99:598–599.

Hebert, P. D. N., M. Y. Stoekle, T. S. Zemlak, and C. M. Francis. 2004. Identification of birds through DNA barcodes. PLoS Biology 2: 1657–1663.

Hellmayr, C. E. 1938. Catalogue of Birds of the Americas. Field Museum of Natural History Publications, Zoology Series, vol. 13, part 11.

Hubbard, J. P. 1974. Geographic variation in the Savannah Sparrows of the inland southwest, Mexico, and Guatemala. Occasional Papers of the Delaware Museum of Natural History, no. 12.

Johnson, N. K. 1982. Retain subspecies—At least for the time being. Auk 99:605–606.

Johnson, N. K., and C. Cicero. 2004. New mitochondrial DNA data affirm the importance of Pleistocene speciation in North American birds. Evolution 58:1122–1130.

Johnson, N. K., J. V. Remsen, Jr., and C. Cicero. 1999. Resolution of the debate over species concepts in ornithology: A new comprehensive species concept. Pages 1470–1482 *in* Acta XXII Congressus Internationalis Ornithologici (N. J. Adams and R. H. Slotow, Eds.). BirdLife South Africa, Johannesburg.

Kerr, K. C. R., M. Y. Stoeckle, C. J. Dove, L. A. Weigt, C. M. Francis, and P. D. N. Hebert. 2007. Comprehensive DNA barcode coverage of North American birds. Molecular Ecology Notes doi:10.1111/j.1471-8286.2006.01670.x

Mayr, E. 1982. Of what use are subspecies? Auk 99:593–595.

Mayr, E., and P. D. Ashlock. 1991. Principles of Systematic Zoology. McGraw-Hill, New York.

McKitrick, M. C., and R. M. Zink. 1988. Species concepts in ornithology. Condor 90:1–14.

Mengel, R. M. 1965. The birds of Kentucky. Ornithological Monographs, no. 3.

Møller, A. P., and J. J. Cuervo. 1998. Speciation and feather ornamentation in birds. Evolution 52:848–858.

Monroe, B. L., Jr. 1982. A modern concept of the subspecies. Auk 99:608–609.

Moritz, C., and C. Cicero. 2004. DNA barcoding: Promise and pitfalls. PLoS Biology 2:1529–1531.

Mumme, R. L., M. L. Galatowitsch, P. G. Jabłoński, T. M. Stawarczyk, and J. P. Cygan. 2006. Evolutionary significance of geographic variation in a plumage-based foraging adaptation: An experimental test in the Slate-throated Redstart (*Myioborus miniatus*). Evolution 60: 1086–1097.

Patten, M. A., and P. Unitt. 2002. Diagnosability versus mean differences of Sage Sparrow subspecies. Auk 119:26–35.

Paynter, R. A., Jr. 1970. Check-list of Birds of the World, vol. 13. Museum of Comparative Zoology, Cambridge, Massachussetts.

Peters, J. L., and L. Griscom. 1938. Geographic variation in the Savannah Sparrow. Bulletin of the Museum of Comparative Zoology 80:443–479.

Phillimore, A. B., and I. P. F. Owens. 2006. Are subspecies useful in evolutionary and conservation biology? Proceedings of the Royal Society of London, Series B 273:1049–1053.

Pyle, P. 1997. Identification Guide to North American Birds. Slate Creek Press, Bolinas, California.

Reed, D. H., and R. Frankham. 2001. How closely correlated are molecular and quantitative measures of genetic variation? A meta-analysis. Evolution 55:1095–1103.

Remsen, J. V., Jr. 2005. Pattern, process, and rigor meet classification. Auk 122:403–413.

Rising, J. D. 2001. Geographic variation in size and shape of Savannah Sparrows (*Passerculus sandwichensis*). Studies in Avian Biology, no. 23.

Rising, J. D., and D. D. Beadle. 1996. A Guide to the Identification and Natural History of the Sparrows of the United States and Canada. Academic Press, San Diego, California.

Rohwer, S. A., and M. S. Johnson 1992. Differences in timing and number of molts for Baltimore and Bullock's orioles: Implications to hybrid fitness and theories of delayed plumage maturation. Condor 95:125–140.

Sol, D., D. G. Stirling, and L. Lefebvre. 2005. Behavioral drive or behavioral inhibition in evolution: Subspecific diversification in Holarctic passerines. Evolution 59:2669–2677.

Storer, R. W. 1982. Subspecies and the study of geographic variation. Auk 99:599–601.

Todd, W. E. C. 1963. Birds of the Labrador Peninsula and Adjacent Areas. University of Toronto Press, Toronto, Ontario.

Townsend, C. W., and G. M. Allen. 1907. Birds of Labrador. Proceedings of the Boston Society of Natural History 33:277–428.

van Rossem, A. J. 1938. Description of twenty-one new races of Fringillidae and Icteridae from Mexico and Guatemala. Bulletin of the British Ornithologists' Club 58:124–138.

van Rossem, A. J. 1947. A synopsis of the Savannah Sparrows of northern Mexico. Condor 49: 97–107.

Wheelwright, N. T., and J. D. Rising. 1993. Savannah Sparrow (*Passerculus sandwichensis*). *In* The Birds of North America, no. 45 (A. Poole and F. Gill, Eds.). Academy of Natural Sciences, Philadelphia, and American Ornithologists' Union, Washington, D.C.

Wiens, J. A. 1982. Forum: Avian subspecies in the 1980's. Auk 99:593.

Wilson, E. O., and W. J. Brown, Jr. 1953. The subspecies concept and its taxonomic application. Systematic Zoology 2:97–111.

Winker, K., D. A. Rocque, T. M. Braile, and C. L. Pruett. 2007. Vainly beating the air: Species-concept debates need not impede progress in science or conservation. Pages 30–44 *in* Festschrift for Ned K. Johnson: Geographic Variation and Evolution in Birds (C. Cicero and J. V. Remsen, Jr., Eds.). Ornithological Monographs, no. 63.

Zink, R. M. 2004. The role of subspecies in obscuring avian biological diversity and misleading conservation policy. Proceedings of the Royal Society of London, Series B 271:561–564.

Zink, R. M., D. L. Dittmann, S. W. Cardiff, and J. D. Rising. 1991. Mitochondrial DNA variation and the taxonomic status of the large-billed Savannah Sparrow. Condor 93:1016–1019.

Zink, R. M., J. D. Rising, S. Mockford, A. G. Horn, J. M. Wright, M. Leonard, and M. C. Westberg. 2005. Mitochondrial DNA variation, species limits, and rapid evolution of plumage coloration and size in the Savannah Sparrow. Condor 107:21–28.

Zusi, R. L. 1982. Infraspecific geographic variation and the subspecies concept. Auk 99:606–608.

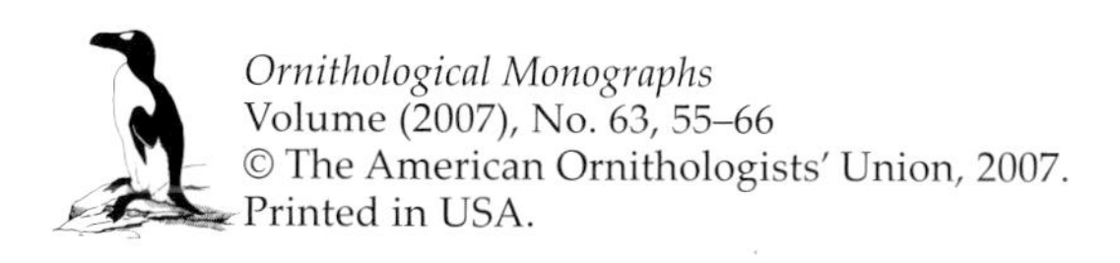
*Ornithological Monographs*
Volume (2007), No. 63, 55–66


CHAPTER 5

# DISTRIBUTIONAL DYNAMICS OF INVASION AND HYBRIDIZATION BY *STRIX* SPP. IN WESTERN NORTH AMERICA

William B. Monahan[1] and Robert J. Hijmans[2]

*Museum of Vertebrate Zoology, 3101 Valley Life Sciences Building, University of California, Berkeley, California 94720, USA*

Abstract.—Understanding how hybridization may affect extant and emerging taxa requires knowledge of the origins, viability, and breeding tendencies of hybrid individuals. We examined the geographic and environmental underpinnings of invasion and hybridization among native North American members of the genus *Strix*. In the early 20th century, the Barred Owl (*S. varia*) started expanding westward through southern Canada. Since 1973, the species has been invading habitats of the Spotted Owl (*S. occidentalis*), usually displacing but occasionally hybridizing with two Spotted Owl subspecies (*S. o. occidentalis* and *S. o. caurina*) to yield viable offspring that often disperse long distances before breeding. Given the high dispersal capabilities of the hybrids, questions remain as to whether hybrid offspring are preferentially colonizing environments that are characteristically different from their natal territories. Further questions surround the proximate origins of the Barred Owl range expansion. We show that the westward range expansion of the Barred Owl was spatiotemporally concomitant with historical increases in summer temperature and that the expansion corridor was positioned at relatively high latitudes because of habitat restrictions. These results provide quantitative support to previous claims that that the Barred Owl range expansion was largely shaped by natural processes. We also show that first filial Barred × Spotted hybrid owls occupy similar environments before and after postnatal dispersal and are randomly distributed both geographically and climatically with respect to parentals. We discuss the climatic mechanisms that may be influencing these dynamics, and conclude with implications for conservation of *Strix* spp. *Received 5 July 2006, accepted 5 February 2007.*

Resumen.—Para entender cómo la hibridación puede afectar a taxones existentes y emergentes, se necesita conocer los orígenes, la viabilidad y las tendencias de apareamiento de los individuos híbridos. En este trabajo, examinamos las bases geográficas y ambientales de la invasión e hibridación entre miembros del genero *Strix*, nativos de Norte América. A principios del siglo 20, *S. varia* comenzó a expandirse desde el oeste hacia el sur de Canadá. Desde 1973, la especie ha invadido habitats de *S. occidentalis*, generalmente desplazándolo, aunque ocasionalmente hibridando con dos de sus subespecies (*S. o. occidentalis* y *S. o. caurina*) y produciendo una progenie viable que normalmente se dispersa a grandes distancias antes de reproducirse. Dada la gran capacidad de dispersión de los híbridos, nos preguntamos si su descendencia coloniza preferentemente ambientes distintos de sus territorios de origen. Cabe también preguntarse por el origen de la expansión del rango de *S. varia*. Nuestros resultados muestran que la expansión de su rango hacia el oeste está relacionada históricamente con un aumento de las temperaturas estivales; además, el corredor de la expansión se localizó a latitudes altas, ésto debido a restricciones en el hábitat. Estos resultados apoyan cuantitativamente las hipótesis de que la expansión del rango de distribución de *S. varia* estuvo determinada por procesos naturales. Además, mostramos que los híbridos descendientes del cruce *S. varia* × *S. occidentalis* ocupan ambientes similares, tanto climática como geográficamente, al de los parentales, presentando, además, una distribución aleatoria. Discutimos los mecanismos climáticos que podrían estar influenciando estas dinámicas y concluimos con implicaciones para la conservación de *Strix*.

---

[1]E-mail: monahan@berkeley.edu
[2]Present address: International Rice Research Institute, DAPO 7777, Manila, Philippines.

Invading species can exert evolutionary influences on resident species through a variety of processes, including physical or niche displacement, demographic breakdown, and hybridization (Mooney and Cleland 2001). Most controversial is the role of hybridization (O'Brien and Mayr 1991, Allendorf et al. 2001). On one hand, hybridization has been associated with many plant and animal extinctions (Rhymer and Simberloff 1996); on the other, it has generated new species (V. Grant 1966, Arnold 1992, Bullini 1994, Rieseberg 1997, Seehausen 2004). Some studies have suggested that hybridization is rare (Dowling and Secor 1997), whereas others have concluded that it is common (Grant and Grant 1992).

Hybridization can act to influence evolutionary trajectories in six major ways, broadly related to whether the causal mechanism is natural or anthropogenic, whether hybrids are fit and fertile, and, if they are fertile, whether they breed with other hybrids or parentals (Arnold and Hodges 1995, Allendorf et al. 2001). Anthropogenically mediated hybridization can result directly from species introductions (e.g., Rhymer et al. 1994) or indirectly via environmental perturbations (e.g., Simons et al. 2001). Unfortunately, it is often difficult in practice to clearly distinguish natural origins from indirect human influences, as is evident with the red wolf (*Canis rufus*; Wayne and Jenks 1991, Wilson et al. 2000). However, such distinctions may be critical, because misclassifications can lead to the failure to protect "natural" biodiversity (Allendorf et al. 2001).

We investigated the geographic and environmental characteristics of invasion and hybridization by *Strix* spp. in western North America. Native to eastern North America, the Barred Owl (*S. varia*) started expanding west from the midwestern United States through southern Canada in the early 1900s (Houston and McGowan 1999). The species first contacted the northern range limit of the Spotted Owl (*S. occidentalis*) in southwestern Canada around 1973 (Taylor and Forsman 1976) and now occurs in western North America from British Columbia south to northern California, largely overlapping the range of the Northern Spotted Owl (*S. o. caurina*) and partially overlapping the range of the California Spotted Owl (*S. o. occidentalis*) (Haig et al. 2004a, Barrowclough et al. 2005). A third well-supported subspecies (Barrowclough and Gutiérrez 1990; Barrowclough et al. 1999; Haig et al. 2001, 2004a), the Mexican Spotted Owl (*S. o. lucida*), remains allopatric with the congeners (Fig. 1). Both *S. o. caurina* and *S. o. lucida* are formally listed as threatened under the U.S. Endangered Species Act (U.S. Fish and Wildlife Service 1990, 1993). Hence, the expansion of the Barred Owl in western North America has important implications for Spotted Owl conservation.

Field observations suggest that Barred Owls, generally the larger and more aggressive of the two species, are in most instances displacing or killing Spotted Owls (Leskiw and Gutiérrez 1998, Kelly et al. 2003, Olson et al. 2005). However, occasionally the species hybridize (Hamer et al. 1994), with almost all pairings involving male Spotted Owls mated to female Barred Owls (Haig et al. 2004b), as predicted from patterns of reverse sexual size-dimorphism (females larger than males; Kelly and Forsman 2004). Apparently, current levels of hybridization are low enough that they do not seriously threaten the conservation status of either species (Haig et al. 2004b, Barrowclough et al. 2005), though hybrids are known to backcross with parentals and questions remain about actual hybrid prevalence.

The origins of the Barred Owl's range expansion remain unknown. A distribution model using the environmental attributes of the species' eastern range accurately predicted the existence of suitable western habitats where the Barred Owl now occurs (Peterson and Robbins 2003). The model predicted extensive geographic overlap between *S. varia* and *S. o. caurina*, as well as partial geographic overlap with *S. o. occidentalis* and *S. o. lucida*. However, the model failed to delineate the Canadian expansion corridor, which suggests that the early phases of the Barred Owl movement were not simply the result of a demographic expansion through suitable habitats. Other studies have concluded that the range expansion was natural in origin but fueled by either adaptation to coniferous forests (Boxall and Stepney 1982) or historical climate change (Johnson 1994). However, the Barred Owl expansion may have been influenced by anthropogenic factors, including changes in forest management practices (Root and Weckstein 1994) or the establishment of wooded riparian areas in the Great Plains (Dark et al. 1998).

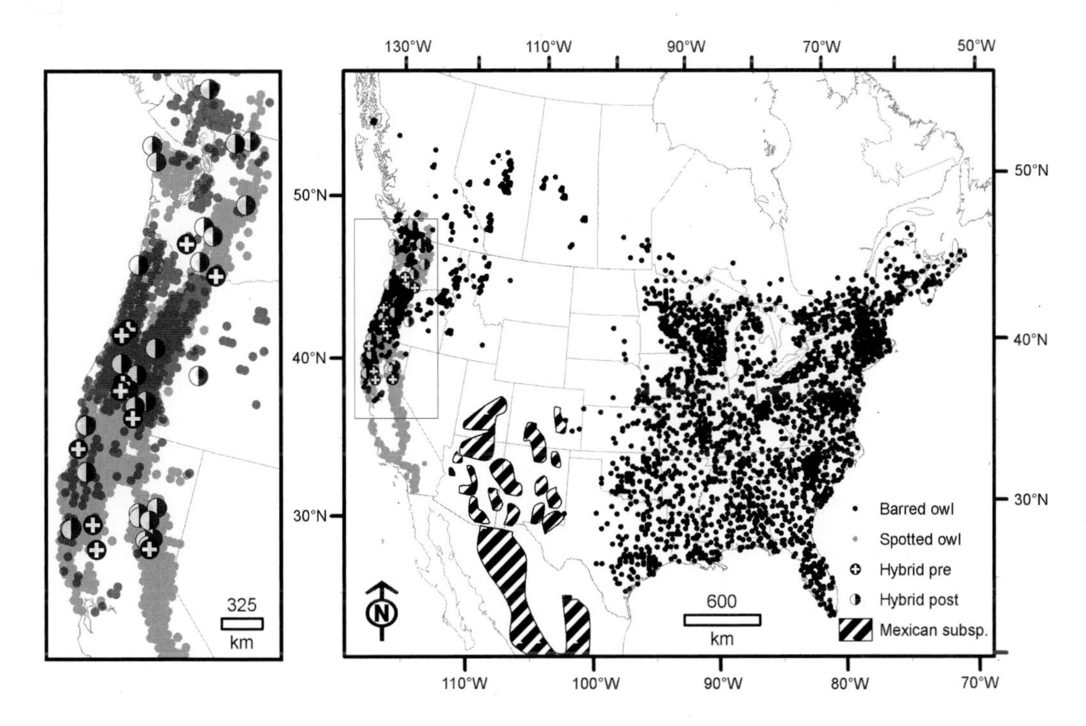


Fig. 1. Observations of Spotted, Barred, and hybrid owls (before and after postnatal dispersal) in North America. The Mexican Spotted Owl (*S. o. lucida*) was not included in the present study (range map obtained from NatureServe; map projection: Albers equal-area conic).

If hybridization in *Strix* represents a largely natural phenomenon tied to the ebb and flow of closely related species' distributions in space and time (Johnson 1994), conservation of these dynamics ultimately requires sound knowledge of the factors that determine where hybrids form, live, and breed. Recent empirical and theoretical work has shown that hybrids possessing extreme phenotypes that enable them to aggregate in physical and novel environmental spaces can become reproductively isolated from parentals and speciate over short evolutionary time-scales (McCarthy et al. 1995, Rieseberg 1997, Buerkle et al. 2000, Lexer et al. 2003, Rieseberg et al. 2003). New phenotypes sometimes appear only after several generations (Rieseberg and Ellstrand 1993). During that period, especially if hybrids remain locally codistributed with parentals, backcrossing can result in the differential shuffling of chromosomal regions between parental taxa (Rieseberg et al. 1995, 1996, 1999). When acted on by selection, such novel genetic variation has been shown to facilitate parental niche expansion (Choler et al. 2004).

Conversely, hybridization in *Strix* could negatively affect parental taxa. Simulations have shown that extinction is a likely outcome of hybridization when reproductive barriers between parentals are weak and the native taxon is both rare and a poor competitor compared with the invader (Wolf et al. 2001). These three factors apply to *Strix* (Burnham et al. 1996, Kelly et al. 2003, Kelly and Forsman 2004, Olson et al. 2005). Extinction is also likely when parental and hybrid taxa occupy similar niches (Wolf et al. 2001). Although Barred and Spotted owls presently occupy similar habitats in western North America, high dispersal capabilities potentially enable the owl hybrids to colonize and breed in areas that are markedly different from their natal environments. Spotted Owls exhibit obligate natal dispersal (Gutiérrez et al. 1995), and we assume that hybrids are similar in this regard. Anecdotally, one hybrid banded as a juvenile was resighted as an adult 292 km away from its natal territory (Forsman et al. 2002). Locations of hybrids before and after postnatal dispersal may thus be distributed either randomly or nonrandomly in geographic or environmental space. Evidence of clustering of hybridization events would suggest an environmentally induced demographic basis for parentals to meet and hybridize, whereas aggregation of adult hybrids would suggest that effective dispersal is nonrandom and capable of promoting ecological isolation. We examine these two possibilities and assess whether the Barred Owl's range expansion was spatiotemporally concomitant with early-20th-century climate change affecting areas with coniferous forests.

## Methods

*Owl data.*—Spotted, Barred, and first-filial-hybrid owl data were compiled from multiple sources, including state and federal (Sauer et al. 2003, Gustafson et al. 2004) wildlife agencies, natural-history museum collections (Burke Museum of Natural History, California Academy of Sciences, Los Angeles County Museum of Natural History, and Museum of Vertebrate Zoology, accessed through ORNIS; see Acknowledgments), and the literature (J. Grant 1966, Reichard 1974, Taylor and Forsman 1976, Boxall and Stepney 1982, Sharp 1989, Dunbar et al. 1991, Hamer et al. 1994, Dark et al. 1998, Kelly 1999, Kelly et al. 2003, Kelly and Forsman 2004). Second-generation hybrids were excluded from the analysis because of potential assignment errors (Haig et al. 2004b, Kelly and Forsman 2004). Non-georeferenced point-occurrence data with locality descriptions were georeferenced using the method described by Wieczorek et al. (2004). Additional localities were extracted from scanned and georeferenced original point-distribution maps (Kelly 1999, Kelly et al. 2003). Spatial uncertainty of georeferenced coordinates was small in relation to the spatial resolution of the analyses. Point data sets were generalized to a 10-minute spatial resolution to reduce the potential of sampling bias affecting our results. The final data set consisted of 959 Spotted Owl localities (672 *S. o. caurina*, 287 *S. o. occidentalis*; we did not consider *S. o. lucida* in the present analysis), 2,902 Barred Owl localities, and 41 hybrid localities, including 11 "juvenile" and 30 "adult" locations (Fig. 1). In all cases, hybrid age categories were determined unambiguously from the average age of owl natal dispersal (September of hatching year; Forsman et al. 2002).

*Origins of the Barred Owl invasion.*—Johnson (1994) hypothesized that the Barred Owl range expansion was facilitated by regional increases in mean summer temperature and precipitation. These changes are believed to have allowed the expansion "corridor" to become climatically more similar to the putative "source" areas. We used mean monthly temperature and precipitation data from June, July, and August (Mitchell and Jones 2005) to estimate changes in summer temperature and precipitation during the period (1901–1970) when the Barred Owl was expanding through southern Canada. To estimate changes in these climate variables, we used means of two 10-year time blocks: 1901–1910 (i.e., "historical," at the beginning of the westward expansion) and 1961–1970 ("modern," when the species reached the Pacific Coast). These periods were long enough to smooth over interannual variation in temperature and precipitation. Barred Owl observations from the expansion corridor (Manitoba, Saskatchewan, Alberta, and British Columbia; $n$ = 133) and from putative source populations (95–85°W and 45–50°N, which included southwestern portions of Ontario and northern portions of Minnesota, Wisconsin, and Michigan; $n$ = 129), were used to assess whether regional patterns of summer warming and increased precipitation caused corridor and source locations to become climatically more similar. We mapped the owl localities on top of layers of summer temperature, precipitation, and vegetation (Latifovic et al. 2002) to assess support for the Barred Owl range expansion being constrained by the presence of coniferous forests (Boxall and Stepney 1982). For purposes of significance testing, and in an effort to reduce problems of non-independence associated with spatial autocorrelation of the data, we iteratively randomly sampled 50 Barred Owl localities (100,000 times) from the source and corridor polygons.

*Distributions of owls in geographic and climatic space.*—We characterized the distributions of Barred Owl, Spotted Owl, and Barred × Spotted owl hybrid offspring in both geographic and climatic space. Nineteen climate variables (Hijmans et al. 2005) broadly summarized temperature and precipitation means, as well as seasonal variability (Table 1). We used principal component analysis (PCA) on the climate data to identify three composite axes explained most of the total sample variation. Statistical significance of the component loadings was determined using a randomization procedure (10,000 iterations) with Bonferroni correction ($\alpha$ = 0.05/57). We then used a second-order point process measure, the $K$ function (Ripley 1976), to determine across different spatial scales whether the owl taxa were randomly or nonrandomly distributed within regions of parental overlap. We compared the $K$ function of the observed number of owl occurrences within distance ($d$) of an arbitrary point of occurrence with the expected number of occurrences, here modeled as a random (Poisson) process where $K(d) = \pi d^2$. Values of $K$ greater than $\pi d^2$ indicate higher observed densities than expected by chance (i.e., clustering), whereas $K$ values less than $\pi d^2$ indicate lower observed densities than expected by chance (i.e., avoidance). Functions of $K$ were computed using latitude and longitude and with the scores from the first three principal components. Geographic simulations were restricted to a minimum convex polygon encompassing all localities where Spotted and Barred owls occur in

TABLE 1. Climate variables used in the principal component (PC) analysis.

| Variable | PC 1 | PC 2 | PC 3 |
|---|---|---|---|
| Annual mean temperature | 0.161 | –0.943* | 0.248 |
| Mean diurnal temperature range | 0.050 | –0.028 | 0.788* |
| Isothermality | 0.853* | –0.176 | 0.389 |
| Temperature seasonality | –0.943* | 0.183 | –0.087 |
| Maximum temperature of warmest month | –0.269 | –0.764* | 0.498 |
| Minimum temperature of coldest month | 0.698* | –0.671* | 0.167 |
| Temperature annual range | –0.903* | 0.229 | 0.136 |
| Mean temperature of wettest quarter | –0.535 | –0.601* | 0.014 |
| Mean temperature of driest quarter | 0.748* | –0.369 | 0.243 |
| Mean temperature of warmest quarter | –0.335 | –0.880* | 0.227 |
| Mean temperature of coldest quarter | 0.595* | –0.757* | 0.240 |
| Annual precipitation | 0.594* | –0.262 | –0.704* |
| Precipitation of wettest month | 0.825* | 0.006 | –0.450 |
| Precipitation of driest month | –0.406 | –0.614* | –0.557 |
| Precipitation seasonality | 0.698* | 0.450 | 0.391 |
| Precipitation of wettest quarter | 0.822* | 0.035 | –0.460 |
| Precipitation of driest quarter | –0.326 | –0.636* | –0.601* |
| Precipitation of warmest quarter | –0.546 | –0.607* | –0.412 |
| Precipitation of coldest quarter | 0.877* | 0.087 | –0.396 |

*Randomization test for significance of component loadings: $P < 0.05/57$.

sympatry (roughly 125–118.5°W and 37.5–51°N). Climate simulations were run for this same invaded region projected into principal-component space, as defined by all two-dimensional combinations of the first three composite axes, which explained 87% of the cumulative percentage trace. Analyses proceeded separately for Spotted, Barred, and hybrid owls, both before and after postnatal dispersal, with 1,000 iterations per simulation.

## RESULTS

*Historical climate change and the Barred Owl invasion.*—North America experienced spatially heterogeneous changes in summer temperature and precipitation during the early to mid-20th century (Fig. 2A, B). Patterns of warming were particularly pronounced in the Barred Owl expansion corridor (Fig. 3A). Areas in the extreme north of the corridor warmed by more than 2°C. Temperature increases in the putative source areas were comparably small in magnitude (Fig. 3A). The mean change in summer temperature for the corridor (+1.1°C) was significantly greater than zero ($t = 2.8$, df = 96, $P < 0.05$), whereas the mean change for the source (+0.5°C) did not deviate significantly from zero ($t = 2.0$, df = 96, $P > 0.05$). Furthermore, the mean increase in summer temperature in the corridor was significantly greater than that in the source ($t = 8.8$, df = 82, $P < 0.001$). Hence, although both the corridor and source localities had summer warming during the early to mid-20th century, the corridor warmed considerably more than the source. At the beginning of the 20th century, the source was 4.0°C warmer than the corridor, but by 1970, the temperature difference had decreased to 3.3°C. The temperature difference between the source in the first decade of the 20th century and the corridor around 1970 was 2.9°C. These warming trends were largely incremental from 1901 through 1970 (Fig. 4).

Contrary to the original hypothesis, both the corridor and source localities became drier during the early to mid-20th century (Fig. 3B). The mean change in summer precipitation for the source was significantly less than zero (–45 mm) ($t = -2.7$, df = 83, $P < 0.05$), whereas mean change for the corridor (–38 mm) did not deviate significantly from zero ($t = -1.3$, df = 81, $P > 0.05$). Furthermore, the mean decrease in summer precipitation in the corridor was not significantly different from that in the source ($t = 0.3$, df = 88, $P > 0.05$). Hence, the Barred Owl range expansion was not spatiotemporally concomitant with historical changes in summer precipitation.

Changes in climate also do not explain why the expansion corridor was positioned at such high latitudes (i.e., "cold" areas). Areas south of the corridor were warmer and would have afforded a more direct route to western North

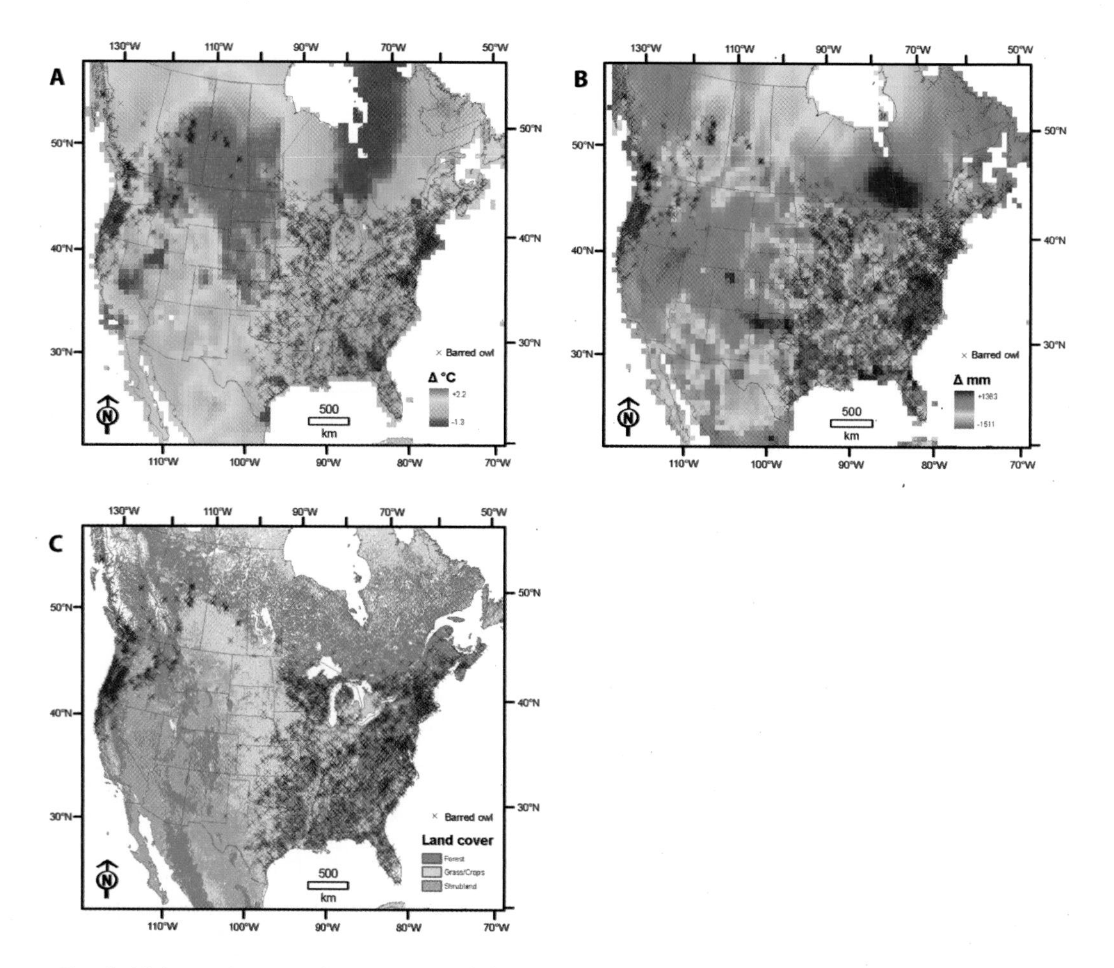


FIG. 2. Major environmental associations of the Barred Owl. (A) Change in mean summer temperature, mean(1961:1970) – mean(1901:1910). (B) Change in mean summer precipitation, mean(1961:1970) – mean(1901:1910). (C) Land cover classification derived from Latifovic et al. (2002). (Map projection: Albers equal-area conic.)

America. However, southern areas lacked the forested habitats that broadly characterize the core eastern range of the Barred Owl (Fig. 2C).

*Geographic and climatic distributions of Barred, Spotted, and hybrid owls.*—Principal component analysis simplified interpretation of the climatic associations of Barred, Spotted, and hybrid owls. The first three principal components accounted for most (87%) of the sample variation: 41% (PC 1), 28% (PC 2), and 18% (PC 3). The first principal component received especially large negative loadings from seasonal and annual measures of temperature variation (Table 1); PC 2 was characterized mostly by large negative temperature loadings; and PC 3 received large positive loadings from mean diurnal temperature range and large negative loadings from annual precipitation and precipitation of the driest quarter.

The first two principal components captured distinct longitudinal gradients separating eastern Barred Owls from all western taxa (Fig. 5). Interpreting these components in light of the original variables, eastern Barred Owls occupied drier and climatically more variable environments than western owls. The second and third principal components failed to identify any additional groupings. Within invaded western regions, hybrids before and after postnatal dispersal were randomly distributed throughout areas of parental sympatry (Fig. 6A). Hybrids at both life stages were also randomly distributed throughout parental climate space as characterized by the first three principal components (Fig. 6B–D).

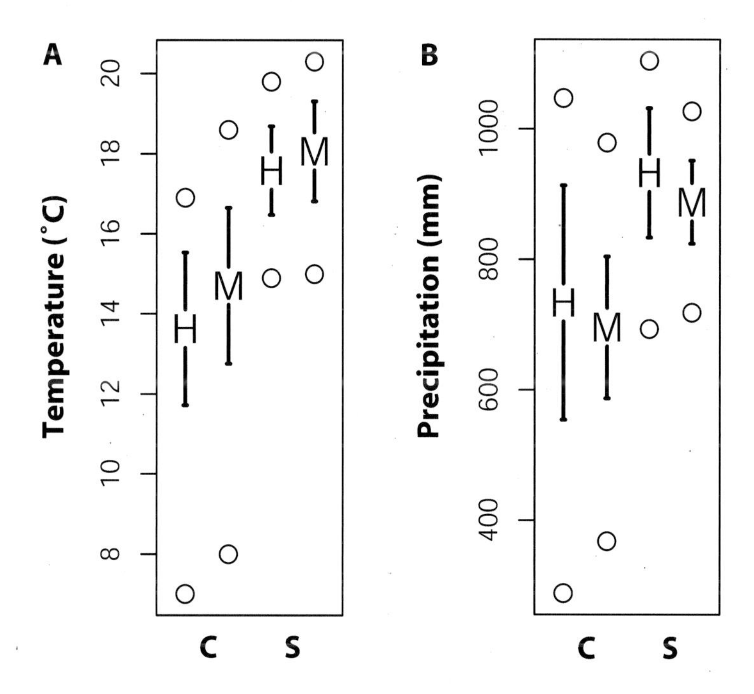


FIG. 3. Change in (A) mean summer temperature and (B) precipitation: Barred Owl expansion corridor (C; $n$ = 133) vs. putative source (S; $n$ = 129) localities (mean ± 1 SD; open circles = extremes; H = "historical," 1901–1910; M = "modern," 1961–1970).

## DISCUSSION

Our results confirm that the Barred Owl range expansion was concomitant with historical increases in mean summer temperature in the region, but not with changes in precipitation (Johnson 1994). Summer is a bioenergetically challenging season for the Barred Owl. Parents must continually provision offspring from spring hatching through late summer or early fall (Mazur and James 2000). Because metabolic energy requirements are inversely related to ambient temperature (Brown et al. 2004), the observation that the corridor warmed significantly more than the source shows that the two portions of the Barred Owl range became energetically more similar as the century progressed. Hence, climatically induced decreases in metabolic energy demand could explain in part how the Barred Owl was able to expand its range through such high latitudes and ultimately colonize western Spotted Owl habitats.

The fact that the expansion corridor traversed southern Canada, which despite summer warming remains colder than most other portions of the species' range, likely stems from habitat restrictions. Boxall and Stepney (1982) speculated that there may have been a recent adaptation by the Barred Owl to coniferous forests. Adaptation or not, a more southern route would have had to traverse grassland and shrubland habitats that, in the absence of large trees with secondary cavities for nesting, are not considered suitable breeding habitats (Mazur and James 2000). In summary, the Barred Owl range expansion is best characterized as "natural" because (1) a distribution model accurately predicted the availability of western habitats where the species now occurs (Peterson and Robbins 2003), (2) the expansion corridor traversed forested areas that are largely representative of the core eastern range of the species, and (3) the timing of the expansion predated the clear onset of anthropogenically mediated climate warming—here taken to be about 1970, because temperature trends after this period cannot be explained by changes in solar radiation (Lean 1997).

Our results also show that juvenile and adult owl hybrids are randomly distributed both geographically and climatically with respect to Spotted and Barred Owl parentals, and that hybrids before and after postnatal dispersal

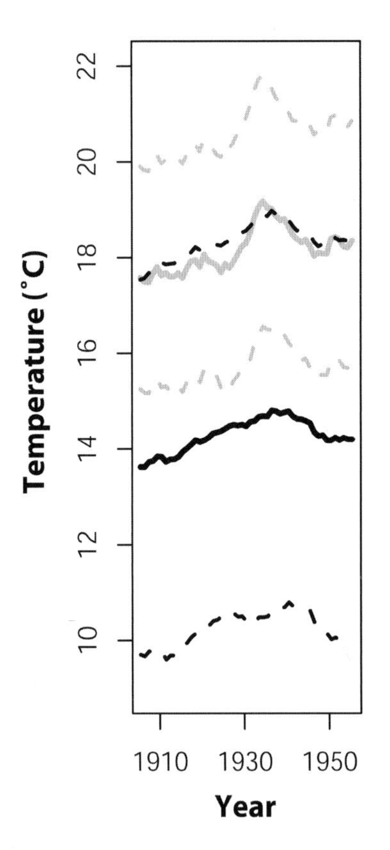


FIG. 4. Time series analysis of change in mean summer temperature (based on a 10-year moving window) in the Barred Owl expansion corridor (black) and in putative source (gray) localities. Solid lines indicate mean values; dashed lines show the 95% confidence intervals.

occupy similar geographic and climatic spaces. These findings suggest that owl hybrids have originated repeatedly in different environments and possess dispersal tendencies that favor parental backcrossing. The observation that hybrids can emerge at low frequencies throughout a large zone of parental overlap is significant, because it suggests opportunities for independent origins of recombinant genotypes in the very beginnings of hybrid establishment (Rieseberg 1997), which in turn increases the probability of transgressive characters arising and quickly becoming fixed (Buerkle et al. 2000). Following hybridization, dispersal can influence hybrid dynamics either by promoting the differential colonization of particular habitats or by ensuring random colonization of mostly parental environments. In the case of the owls, dispersal has allowed hybrids to remain randomly

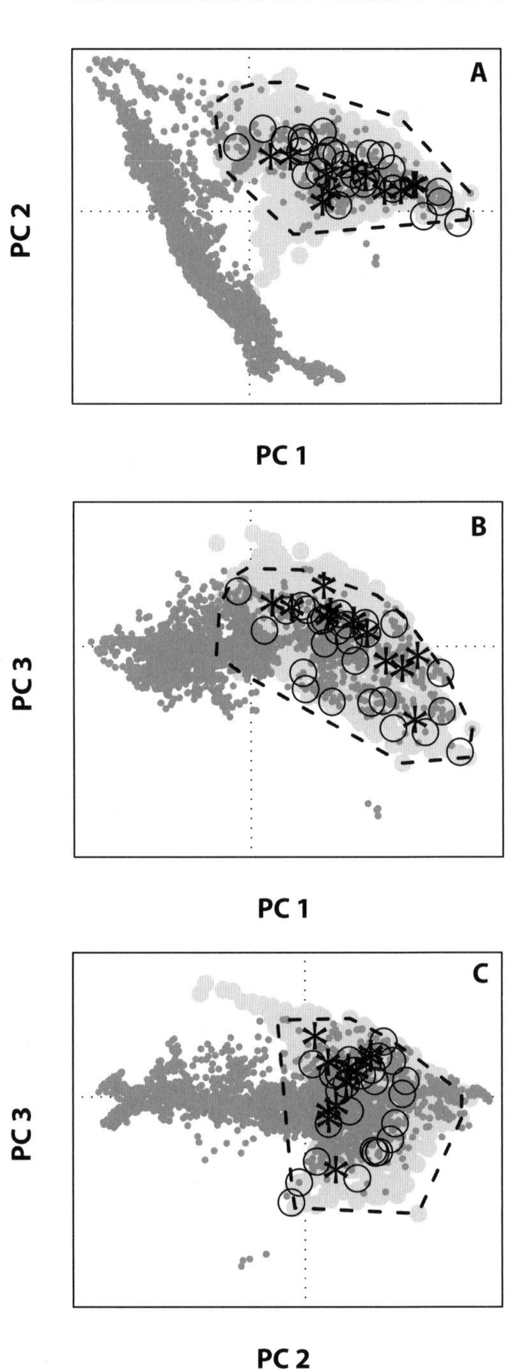


FIG. 5. Bivariate plots of component scores derived from PCA on climate variables. Dark gray points identify *S. varia*, light gray circles *S. occidentalis*, asterisks hybrids before postnatal dispersal, and open black circles hybrids after postnatal dispersal. Dashed black line delineates the invaded geographic space projected into component space.

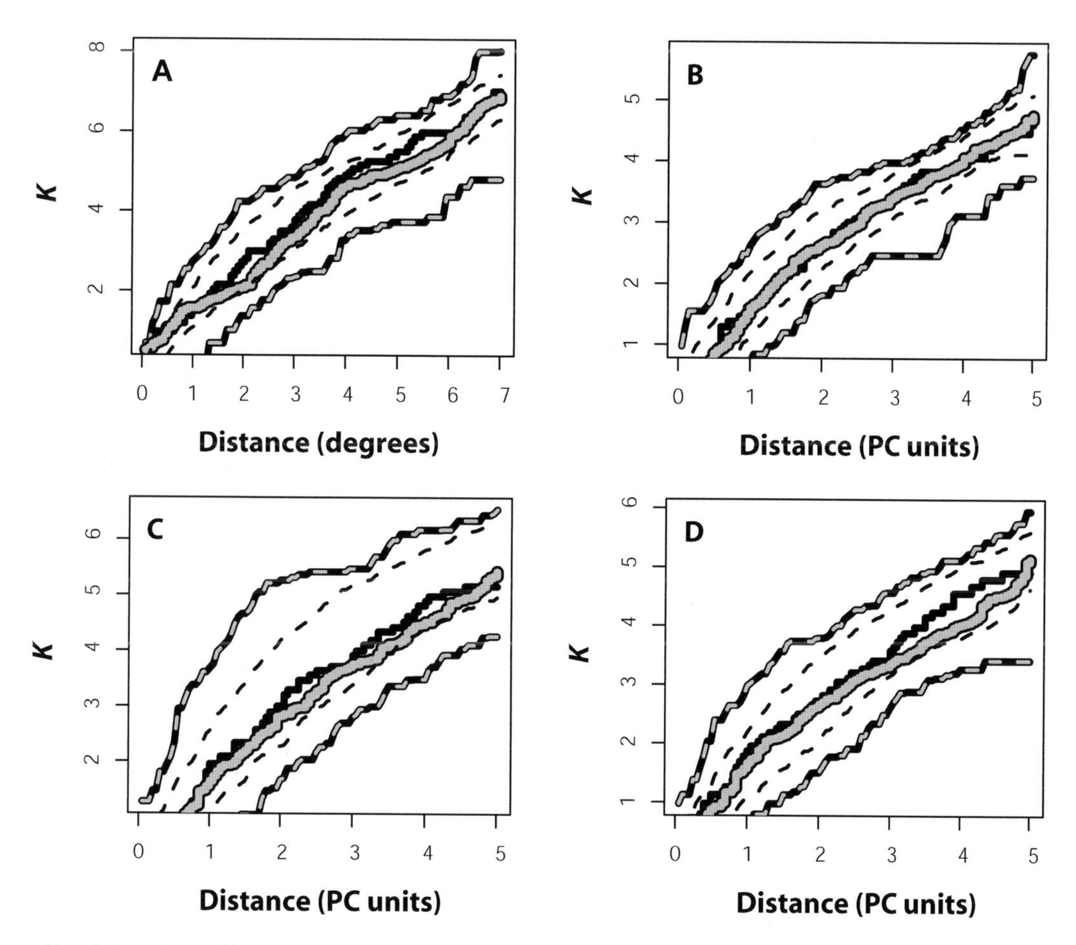


Fig. 6. Functions of $K$ (Ripley 1976) describing distributions of hybrids before (gray; $n$ = 11) and after (black; $n$ = 30) postnatal dispersal (A) in invaded geographic space and (B–D) in multivariate climatic space as characterized by the first three principal components in the PCA: PC 2 vs. PC 1 (B), PC 3 vs. PC 1 (C), and PC 3 vs. PC 2 (D). Dashed lines identify minimum and maximum bounds of random expectations. Because in all cases the observed $K$ values are bounded by the null $K$ values, the results suggest that hybrid owls at both life stages are randomly distributed with respect to parentals.

distributed as breeding adults. However, because we were unable to follow individuals through development, what remains unknown is whether hybrids are tracking natal environments or truly dispersing randomly with respect to the geography and climate of the parentals.

Our results contribute two major points to the discussion of how hybridization should be considered in the context of conservation of *Strix* spp. First, the proximate origins of interspecific hybridization are, at large spatial scales, explained in part by climate and historical changes in climate. Because the Barred Owl range expansion tracked forested habitats and occurred during a period when warming trends could be attributed to natural forces without invoking anthropogenic influences, we suggest that recent opportunities for Barred and Spotted owl hybridization stem in large part from natural processes. Second, Barred, Spotted, and first filial Barred × Spotted owl hybrids are shown to exhibit high levels of geographic and niche overlap, at least as measured according to climate. This spatial arrangement promotes parental backcrossing over assortative hybrid mating. In such cases of weak niche or habitat differentiation, one taxon ultimately is expected to replace the other two (Wolf et al. 2001).

Given the recency of the Barred Owl invasion, conservation of *Strix* spp. in the short term will still depend largely on whether critical Spotted Owl habitats continue to persist (Gutiérrez 1994, Gutiérrez et al. 1995). Because the issue of habitat availability is of fundamental conservation concern to all species, it is perhaps not surprising that most endangered-species legislation has focused more on mitigating threats posed by habitat loss than on addressing invasion and hybridization. However, the challenge posed by hybridization will need to be addressed increasingly in the coming decades, as climate change and other human-mediated invasions continue to pave the way for new species interactions (Mooney and Cleland 2001). Certainly, genetic approaches to identifying hybrid influences will be critical. We emphasize that spatial considerations of parentals and hybrids also can help predict future dynamics.

## Acknowledgments

We thank G. Gould and R. Gutiérrez for furnishing California Department of Fish and Game data, U.S. Geological Survey Patuxent Wildlife Research Center for providing access to Breeding Bird Survey and Bird Banding Lab data, and the Climatic Research Unit for facilitating use of the most recent release of time-series temperature data. We are also indebted to ORNIS (ornisnet.org) and the Burke Museum of Natural History, California Academy of Sciences, Los Angeles County Museum of Natural History, and Museum of Vertebrate Zoology for providing specimen locality data. Comments from C. Moritz, W. Koenig, T. Peterson, R. Pereira, and an anonymous reviewer improved the manuscript. W.B.M. received graduate fellowship support from the National Science Foundation. Most importantly, W.B.M. wishes to thank Ned K. Johnson for his mentorship during graduate school.

## Literature Cited

Allendorf, F. W., R. F. Leary, P. Spruell, and J. K. Wenburg. 2001. The problems with hybrids: Setting conservation guidelines. Trends in Ecology and Evolution 16:613–622.

Arnold, M. L. 1992. Natural hybridization as an evolutionary process. Annual Review of Ecology and Systematics 23:237–261.

Arnold, M. L., and S. A. Hodges. 1995. Are natural hybrids fit or unfit relative to their parents? Trends in Ecology and Evolution 10:67–71.

Barrowclough, G. F., J. G. Groth, L. A. Mertz, and R. J. Gutiérrez. 2005. Genetic structure, introgression, and a narrow hybrid zone between Northern and California spotted owls (*Strix occidentalis*). Molecular Ecology 14:1109–1120.

Barrowclough, G. F., and R. J. Gutiérrez. 1990. Genetic variation and differentiation in the Spotted Owl (*Strix occidentalis*). Auk 107: 737–744.

Barrowclough, G. F., R. J. Gutiérrez, and J. G. Groth. 1999. Phylogeography of Spotted Owl (*Strix occidentalis*) populations based on mitochondrial DNA sequences: Gene flow, genetic structure, and a novel biogeographic pattern. Evolution 53:919–931.

Boxall, P. C., and P. H. R. Stepney. 1982. The distribution and status of the Barred Owl in Alberta. Canadian Field-Naturalist 96:46–50.

Brown, J. H., J. F. Gillooly, A. P. Allen, V. M. Savage, and G. B. West. 2004. Toward a metabolic theory of ecology. Ecology 85:1771–1789.

Buerkle, C. A., R. J. Morris, M. A. Asmussen, and L. H. Rieseberg. 2000. The likelihood of homoploid hybrid speciation. Heredity 84:441–451.

Bullini, L. 1994. Origin and evolution of animal hybrid species. Trends in Ecology and Evolution 9:422–426.

Burnham, K. P., D. R. Anderson, and G. C. White. 1996. Meta-analysis of vital rates of the Northern Spotted Owl. Pages 92–101 *in* Demography of the Northern Spotted Owl (E. D. Forsman, S. DeStefano, M. G. Raphael, and R. J. Gutiérrez, Eds.). Studies in Avian Biology, no. 17.

Choler, P., B. Erschbamer, A. Tribsch, L. Gielly, and P. Taberlet. 2004. Genetic introgression as a potential to widen a species' niche: Insights from alpine *Carex curvula*. Proceedings of the National Academy of Sciences USA 101: 171–176.

Dark, S. J., R. J. Gutiérrez, and G. I. Gould. 1998. The Barred Owl (*Strix varia*) invasion in California. Auk 115:50–56.

Dowling, T. E., and C. L. Secor. 1997. The role of hybridization and introgression in the diversification of animals. Annual Review of Ecology and Systematics 28:593–619.

Dunbar, D. L., B. P. Booth, E. D. Forsman, A. E. Hetherington, and D. J. Wilson. 1991. Status of the Spotted Owl, *Strix occidentalis*, and Barred Owl, *Strix varia*, in southwestern British Columbia. Canadian Field-Naturalist 105:464–468.

Forsman, E. D., R. G. Anthony, J. A. Reid, P. J. Loschl, S. G. Sovern, M. Taylor, B. L. Biswell, A. Ellingson, E. C. Meslow, G. S. Miller, and others. 2002. Natal and breeding dispersal of Northern Spotted Owls. Wildlife Monographs, no. 149.

Grant, J. 1966. The Barred Owl in British Columbia. Murrelet 47:39–45.

Grant, P. R., and B. R. Grant. 1992. Hybridization of bird species. Science 256:193–197.

GRANT, V. 1966. The origin of a new species of *Gilia* in a hybridization experiment. Genetics 54: 1189–1199.

GUSTAFSON, M. E., J. HILDENBRAND, AND L. METRAS. 2004. The North American Bird Banding Manual, version 1.0. [Online.] U.S. Geological Survey Patuxent Wildlife Research Center, Laurel, Maryland. Available at www.pwrc.usgs.gov/BBL/manual/manual.htm.

GUTIÉRREZ, R. J. 1994. Changes in the distribution and abundance of Spotted Owls during the past century. Pages 293–300 *in* A Century of Avifaunal Change in Western North America (J. R. Jehl, Jr., and N. K. Johnson, Eds.). Studies in Avian Biology, no. 15.

GUTIÉRREZ, R. J., A. B. FRANKLIN, AND W. S. LAHAYE. 1995. Spotted Owl (*Strix occidentalis*). *In* The Birds of North America, no. 179 (A. Poole and F. Gill, Eds.). Academy of Natural Sciences, Philadelphia, and American Ornithologists' Union, Washington, D.C.

HAIG, S. M., T. D. MULLINS, AND E. D. FORSMAN. 2004a. Subspecific relationships and genetic structure in the Spotted Owl. Conservation Genetics 5:683–705.

HAIG, S. M., T. D. MULLINS, E. D. FORSMAN, P. W. TRAIL, AND L. WENNERBERG. 2004b. Genetic identification of Spotted Owls, Barred Owls, and their hybrids: Legal implications of hybrid identity. Conservation Biology 18: 1347–1357.

HAIG, S. M., R. S. WAGNER, E. D. FORSMAN, AND T. D. MULLINS. 2001. Geographic variation and genetic structure in Spotted Owls. Conservation Genetics 2:25–40.

HAMER, T. E., E. D. FORSMAN, A. D. FUCHS, AND M. L. WALTERS. 1994. Hybridization between Barred and Spotted owls. Auk 11:487–492.

HIJMANS, R. J., S. CAMERON, J. PARRA, P. G. JONES, AND A. JARVIS. 2005. Very high resolution interpolated climate surfaces for global land areas. International Journal of Climatology 25: 1965–1978.

HOUSTON, C. S., AND K. J. MCGOWAN. 1999. The westward spread of the Barred Owl. Blue Jay 57:191–195.

JOHNSON, N. K. 1994. Pioneering and natural expansion of breeding distributions in western North American birds. Pages 27–44 *in* A Century of Avifaunal Change in Western North America (J. R. Jehl, Jr., and N. K. Johnson, Eds.). Studies in Avian Biology, no. 15.

KELLY, E. G. 1999. The range expansion of the northern Barred Owl: An evaluation of the impact on Spotted Owls. M.S. thesis, Oregon State University, Corvallis.

KELLY, E. G., AND E. D. FORSMAN. 2004. Recent records of hybridization between Barred Owls (*Strix varia*) and Northern Spotted Owls (*S. occidentalis caurina*). Auk 121:806–810.

KELLY, E. G., E. D. FORSMAN, AND R. G. ANTHONY. 2003. Are Barred Owls displacing Spotted Owls? Condor 105:45–53.

LATIFOVIC, R., Z.-L. ZHU, J. CIHLAR, AND C. GIRI. 2002. Land cover of North America 2000. Natural Resources Canada, Canada Center for Remote Sensing, and U.S. Geological Service EROS Data Center.

LEAN, J. 1997. The Sun's radiation and its relevance for Earth. Annual Review of Astronomy and Astrophysics 35:33–67.

LESKIW, T., AND R. J. GUTIÉRREZ. 1998. Possible predation of a Spotted Owl by a Barred Owl. Western Birds 29:225–226.

LEXER, C., M. E. WELCH, O. RAYMOND, AND L. H. RIESEBERG. 2003. The origin of ecological divergence in *Helianthus paradoxus* (Asteraceae): Selection on transgressive characters in a novel hybrid habitat. Evolution 57: 1989–2000.

MAZUR, K. M., AND P. C. JAMES. 2000. Barred Owl (*Strix varia*). *In* The Birds of North America, no. 508 (A. Poole and F. Gill, Eds.). Birds of North America, Philadelphia.

MCCARTHY, E. M., M. A. ASMUSSEN, AND W. W. ANDERSON. 1995. A theoretical assessment of recombinational speciation. Heredity 74: 502–509.

MITCHELL, T. D., AND P. D. JONES. 2005. An improved method of constructing a database of monthly climate observations and associated high-resolution grids. International Journal of Climatology 25:693–712.

MOONEY, H. A., AND E. E. CLELAND. 2001. The evolutionary impact of invasive species. Proceedings of the National Academy of Sciences USA 98: 5446–5451.

O'BRIEN, S. J., AND E. MAYR. 1991. Bureaucratic mischief: Recognizing endangered species and subspecies. Science 251:1187–1188.

OLSON, G. S., R. G. ANTHONY, E. D. FORSMAN, S. H. ACKERS, P. J. LOSCHL, J. A. REID, K. M. DUGGER, E. M. GLENN, AND W. J. RIPPLE. 2005. Modeling of site occupancy dynamics for Northern Spotted Owls, with emphasis on the effects of Barred Owls. Journal of Wildlife Management 69:918–932.

PETERSON, A. T., AND C. R. ROBBINS. 2003. Using ecological niche modeling to predict Barred Owl invasions with implications for Spotted Owl conservation. Conservation Biology 17:1161–1165.

REICHARD, T. A. 1974. Barred Owl sightings in Washington. Western Birds 5:138–140.

RHYMER, J. M., AND D. SIMBERLOFF. 1996. Extinction by hybridization and introgression. Annual Review of Ecology and Systematics 27:83–109.

Rhymer, J. M., M. J. Williams, and M. J. Braun. 1994. Mitochondrial analysis of gene flow between New Zealand Mallards (*Anas platyrhynchos*) and Grey Ducks (*A. superciliosa*). Auk 111:970–978.

Rieseberg, L. H. 1997. Hybrid origins of plant species. Annual Review of Ecology and Systematics 28:359–389.

Rieseberg, L. H., and N. C. Ellstrand. 1993. What can morphological and molecular markers tell us about plant hybridization? Critical Reviews in Plant Sciences 12:213–241.

Rieseberg, L. H., O. Raymond, D. M. Rosenthal, Z. Lai, K. Livingstone, T. Nakazato, J. L. Durphy, A. E. Schwarzbach, L. A. Donovan, and C. Lexer. 2003. Major ecological transitions in wild sunflowers facilitated by hybridization. Science 301:1211–1216.

Rieseberg, L. H., B. Sinervo, C. R. Linder, M. Ungerer, and D. M. Arias. 1996. Role of gene interactions in hybrid speciation: Evidence from ancient and experimental hybrids. Science 272:741–745.

Rieseberg, L. H., C. Van Fossen, and A. M. Desrochers. 1995. Hybrid speciation accompanied by genomic reorganization in wild sunflowers. Nature 375:313–316.

Rieseberg, L. H., J. Whitton, and K. Gardner. 1999. Hybrid zones and the genetic architecture of a barrier to gene flow between two sunflower species. Genetics 152:713–727.

Ripley, B. D. 1976. The second-order analysis of stationary point processes. Journal of Applied Probability 13:255–266.

Root, T. L., and J. D. Weckstein. 1994. Changes in distribution patterns of select wintering North American birds from 1901 to 1989. Pages 191–201 *in* A Century of Avifaunal Change in Western North America (J. R. Jehl, Jr., and N. K. Johnson, Eds.). Studies in Avian Biology, no. 15.

Sauer, J. R., J. E. Hines, and J. Fallon. 2003. The North American Breeding Bird Survey, results and analysis 1966–2002, version 2003.1. [Online.] U.S. Geological Survey Patuxent Wildlife Research Center, Laurel, Maryland. Available at www.mbr-pwrc.usgs.gov/bbs/bbs.html.

Seehausen, O. 2004. Hybridization and adaptive radiation. Trends in Ecology and Evolution 19: 198–207.

Sharp, D. U. 1989. Range extension of the Barred Owl in western Washington and first breeding record on the Olympic Peninsula. Journal of Raptor Research 23:179–180.

Simons, A. M., R. M. Wood, L. S. Heath, B. R. Kuhajda, and R. L. Mayden. 2001. Phylogenetics of *Scaphirhynchus* based on mitochondrial DNA sequences. Transactions of the American Fisheries Society 130:359–366.

Taylor, A. L., and E. D. Forsman. 1976. Recent range extensions of the Barred Owl in western North America, including the first records for Oregon. Condor 78:560–561.

U.S. Fish and Wildlife Service. 1990. Endangered and threatened wildlife and plants: Determination of threatened status for the Northern Spotted Owl. Federal Register 55:26114–26194.

U.S. Fish and Wildlife Service. 1993. Final rule to list the Mexican Spotted Owl as a threatened species. Federal Register 58:14248–14271.

Wayne, R. K., and S. M. Jenks. 1991. Mitochondrial DNA analysis implying extensive hybridization of the endangered red wolf *Canis rufus*. Nature 351:565–568.

Wieczorek, J., Q. Guo, and R. J. Hijmans. 2004. The point-radius method for georeferencing locality descriptions and calculating associated uncertainty. International Journal of Geographic Information Science 18:745–767.

Wilson, P. J., S. Grewal, I. D. Lawford, J. N. M. Heal, A. G. Granacki, D. Pennock, J. B. Theberge, M. T. Theberge, D. R. Voigt, W. Waddell, and others. 2000. DNA profiles of the eastern Canadian wolf and the red wolf provide evidence for a common evolutionary history independent of the gray wolf. Canadian Journal of Zoology 78:2156–2166.

Wolf, D. E., N. Takebayashi, and L. H. Rieseberg. 2001. Predicting the risk of extinction through hybridization. Conservation Biology 15: 1039–1053.

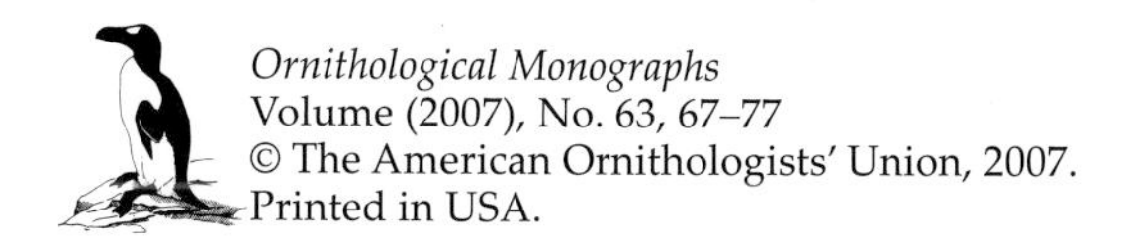

*Ornithological Monographs*
Volume (2007), No. 63, 67–77


CHAPTER 6

# DIVERGENCE BETWEEN SUBSPECIES GROUPS OF SWAINSON'S THRUSH (*CATHARUS USTULATUS USTULATUS* AND *C. U. SWAINSONI*)

Kristen Ruegg[1]

*Museum of Vertebrate Zoology, Department of Integrative Biology, University of California, Berkeley, California 94720, USA*

Abstract.—Swainson's Thrush (*Catharus ustulatus*) is a long-distance Nearctic–Neotropical migrant that includes two major subspecies groups: the russet-backed group (*C. u. ustulatus*) of the Pacific Coast and the olive-backed inland group (*C. u. swainsoni*) (American Ornithologists' Union [AOU] 1998, Evans Mack and Yong 2000). The two groups are most easily distinguished by differences in plumage characteristics, breeding and wintering location, and some vocalizations (Evans Mack and Yong 2000). Historical controversy over the taxonomic treatment of the *ustulatus* and *swainsoni* groups suggests that they have previously been considered on the border between subspecies and recently diverged sister species (reviewed in Bond 1963). Several authors emphasize the differences between the groups in breeding habitat and the lack of intergradation in regions where they co-occur (Bent 1949, Phillips 1991). Other authors emphasize their similarities, citing extensive intergradation along the eastern slope of the Sierra Nevada (Grinnell and Miller 1944). In the past decade, a number of studies have brought further clarity to the extent of genetic, behavioral, ecological, and acoustic divergence between the *ustulatus* and *swainsoni* subspecies groups. Here, I review these recent advances in our knowledge, identify future research questions, and discuss potential implications of divergence on the taxonomic treatment of the two groups according to the AOU's guidelines for naming species (AOU 1998, Johnson et al. 1999). *Received 25 July 2006, accepted 9 February 2007.*

Resumen.—*Catharus ustulatus* es un migrador de larga distancia, Neártico–Neotropical. Incluye dos grupos importantes de subespecies: en la costa del Pacífico, el grupo de "espalda café-rojizo", *C. u. ustulatus*, y, en el interior, el grupo de "espalda olivácea", *C. u. swainsoni* (American Ornithologists' Union 1998, Evans Mack and Yong 2000). Ambos grupos se diferencian fácilmente por las características del plumaje, sitios de reproducción y área de invernada, además de vocalizaciones (Evans Mack and Yong 2000). La controversia sobre el tratamiento taxonómico de los grupos *ustulatus* y *swainsoni*, sugiere que han sido previamente considerados en el límite entre subespecies y especies hermanas que recientemente divergieron (revisado en Bond 1963). Varios autores enfatizan las diferencias entre los grupos en áreas de reproducción y la falta de intergradación en las regiones en que coexisten (Bent 1949, Phillips 1991). Otros autores subrayan sus similitudes, aludiendo además a una extensiva intergradación en la vertiente este de la Sierra Nevada (Grinnell and Miller 1944). En la década pasada, varios estudios clarificaron los niveles de divergencia genética, de comportamiento, acústicas y ecológica entre los grupos de subespecie *ustulatus* y *swainsoni*. En este estudio, reviso los trabajos anteriores, identifico futuras preguntas de investigación, y discuto las implicaciones taxonómicas para ambos grupos de acuerdo con las reglas establecidas por la Unión de Ornitólogos Americanos (American Ornithologists' Union 1998, Johnson et al. 1999).

## Evolutionary History

Recent genetic data have helped shed light on the evolutionary history within the genus *Catharus* as well as on intraspecific variation within the Swainson's Thrush (*C. ustulatus*). Two independently derived molecular phylogenies of the genus *Catharus* using different loci suggest that the Swainson's Thrush is the oldest lineage in the genus and is sister to all other species of *Catharus*, including several resident and short-distance migratory species from the Neotropics (Outlaw et al. 2003, Winker and

[1]E-mail: kruegg@berkeley.edu

Pruett 2006). The five morphologically similar long-distance migratory species within the genus *Catharus* do not form a monophyletic clade, which indicates that migratory behavior has evolved multiple times within the genus and that the morphology of the North American long-distance migratory species exhibits strong evolutionary convergence (Winker and Pruett 2006). Outlaw et al. (2003) suggested that the Swainson's Thrush diverged from other *Catharus* species ~4 mya, and the combined data sets indicate that species of *Catharus* most likely diverged somewhere in the Neotropics.

An analysis of population structure within the Swainson's Thrush (Ruegg and Smith 2002) using a rapidly evolving region of the mitochondrial genome, the control region, focused on genetic differentiation between the two major subspecies groups (Fig. 1): the russet-backed group of the Pacific Coast (*C. u. ustulatus*) and the olive-backed inland group (*C. u. swainsoni*) (American Ornithologists' Union [AOU] 1998, Evans Mack and Yong 2000). These data showed that divergence between the subspecies groups is lower than reported levels of divergence between many sister species, but similar to estimates of divergence between well-differentiated subspecies of birds. Individuals of *ustulatus* and *swainsoni* from across the breeding range are separated by five diagnostic mutations with a net sequence divergence of 0.7% (Fig. 2; Ruegg and Smith 2002). By comparison, Johnson and Cicero (2004) found that estimates of mitochondrial DNA (mtDNA) sequence divergence in 39 pairs of avian sister species ranged from 0.0% to 8.2%, with an average divergence of 1.9%. Alternatively, control-region divergence between some well-differentiated subspecies ranges from 0.07% to 4.8%: 0.07% between Brewer's Sparrow (*Spizella breweri breweri*) and Timberline Sparrow (*S. b. taverneri*) subspecies

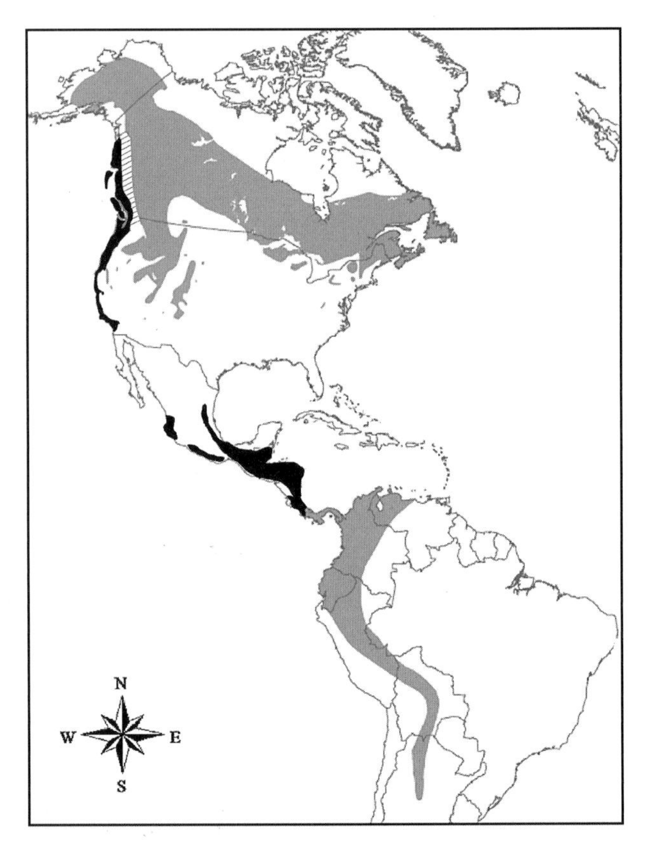


FIG. 1. Breeding and wintering ranges of coastal *ustulatus* (black) and inland *swainsoni* (gray) subspecies groups and potential contact zones (black and white stripes) in Swainson's Thrush (based on a range map from the Cornell Laboratory of Ornithology, with range data by Nature-Serve).

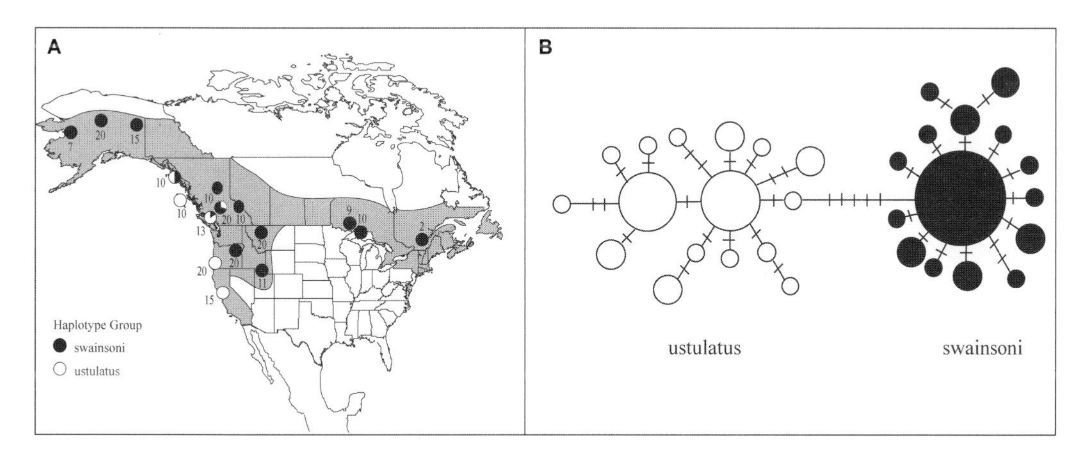


FIG. 2. Summary of genetic analysis based on mtDNA control-region haplotypes from Ruegg and Smith (2002). (A) Frequency of coastal *ustulatus* (white circles) and inland *swainsoni* (black circles) haplotypes in 17 breeding populations. (B) Haplotype network with bars across branches indicating single nucleotide changes. Four sizes of circle are used to represent the number of individuals sharing the same haplotype (smallest, 1 individual; medium-small, 2–4 individuals; medium-large, 5–11 individuals; largest, 44 individuals).

(Klicka et al. 1999), 0.1% between Bluethroat (*Luscinia svecica*) subspecies (Questiau et al. 1998), and 1.7–4.8% between seven subspecies of Common Chaffinch (*Fringilla coelebs*; Marshall and Baker 1998).

Nuclear genetic distances between *ustulatus* and *swainsoni* populations from the Pacific Northwest of North America (two *ustulatus*, two *swainsoni*, and one mixed population) are lower than microsatellite genetic distances between other designated species but within the range of genetic distances between other subspecies and populations. Microsatellite-based genetic distances, in which an $F_{ST}$ of 1 represents no gene flow and an $F_{ST}$ of 0 represents complete mixing, ranged from 0.018 to 0.043 between pure *ustulatus* and *swainsoni* populations (Ruegg et al. 2006b). In comparison, $F_{ST}$-based estimates of genetic distance among some species of birds range from 0.054 (Shy Albatross [*Thalassarche cauta*] and White-capped Albatross [*T. steadi*]; Abbott and Double 2003) to 0.069 (five species of gulls [*Larus* spp.]; Crochet et al. 2003), whereas $F_{ST}$-based genetic distances among some well-differentiated subspecies and populations range from 0.014 to 0.075 (0.014 for populations of Yellow Warblers [*Dendroica petechia*] from across North America [Gibbs et al. 2000]; 0.018 among populations of Chestnut-backed Chickadees [*Poecile rufenscens*; Burg et al. 2006]; and 0.075 among subspecies of North American Steller's Jays [*Cyanocitta stelleri*; Burg et al. 2005]). In conclusion, mitochondrial and nuclear genetic divergence between the *ustulatus* and *swainsoni* subspecies groups is somewhat lower than estimates of genetic divergence between some other avian sister species, but is within the range of genetic divergence between other well-defined subspecies of birds.

It is generally accepted that divergence between many well-differentiated subspecies and recently diverged sister species of birds can be attributed to geographic isolation during the mid- to late Pleistocene (Avise and Walker 1998, Johnson and Cicero 2004, Lovette 2005). Paleoecological data suggest that suitable songbird habitat was present in the east and west, south of the glacier's edge, and that the center of the country was occupied by tundra and desert that likely would have been inhospitable to a songbird (Pielou 1991). Genetic data combined with climate models of the distribution of the *ustulatus* and *swainsoni* subspecies groups at the last glacial maximum (LGM) are concordant with paleoecological data and suggest that the most likely distribution of populations at the LGM would have been western coastal and southeastern regions (Ruegg et al. 2006a; Fig. 3). In light of the hypothesized distribution at the LGM, regions of parapatry between the *ustulatus* and *swainsoni* subspecies groups likely represent regions of secondary contact following postglacial range expansions.

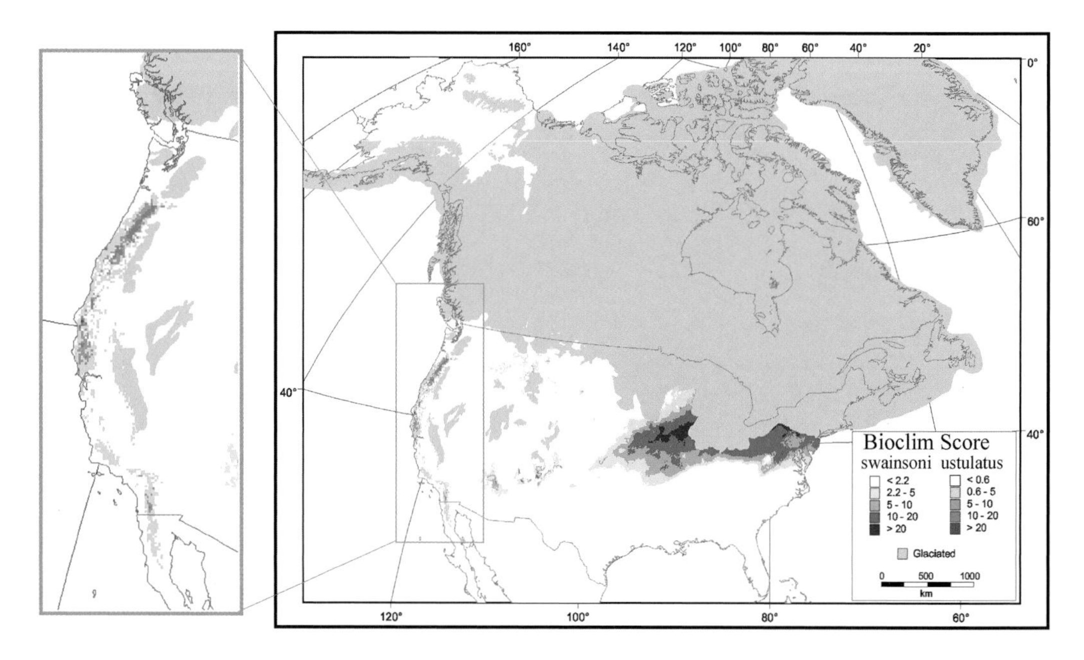


FIG. 3. Model of the distribution of coastal and inland populations at the last glacial maximum, based on climate data (adapted from Ruegg et al. 2006a).

## MIGRATORY ROUTES

Genetic, banding recapture, and subspecies distributional data confirm that the *ustulatus* and *swainsoni* subspecies groups of Swainson's Thrush follow distinct migratory pathways and overwinter in largely allopatric locations (Ramos and Warner 1980, Phillips 1991, Ruegg and Smith 2002). Coastal *ustulatus* populations migrate along a western route to wintering locations in southern Mexico and Central America, whereas inland *swainsoni* populations follow an eastern migratory pathway to wintering locations in Panama and South America (Fig. 1). Distributional records indicate that *ustulatus* populations follow the same migratory route for both fall and spring migration, but that *swainsoni* populations follow different migratory routes in spring and fall. During spring migration, *swainsoni* populations are found throughout the central United States, whereas during fall migration, *swainsoni* populations shift their primary migration route eastward (Evans Mack and Yong 2000).

Northwestern *swainsoni* breeding populations do not follow the shortest migratory route to the wintering grounds. Instead, banding recapture and genetic data indicate that individuals migrate along an eastern route toward ancestral breeding grounds in the east before migrating south to the wintering grounds (Ruegg and Smith 2002). Ruegg and Smith (2002) suggested that the migratory pathway of northwestern *swainsoni* breeding populations is an artifact of their postglacial range expansion and noted that it may not represent the most efficient route to the wintering grounds. Two lines of evidence support this hypothesis: (1) the migratory pathway of northwestern *swainsoni* populations mirrors the postglacial expansion of the boreal forest at the end of the LGM (Ruegg and Smith 2002), and (2) genetic signatures of population expansion and climatic models of the change in distribution since the LGM both support the idea that the *ustulatus* and *swainsoni* subspecies groups expanded out of eastern and western regions following the recession of ice sheets from North America (Ruegg and Smith 2002, Ruegg et al. 2006a). An alternative (and not mutually exclusive) explanation to this historical hypothesis is that the current migratory pathway of northwestern *swainsoni* populations is adaptive because of favorable ecological conditions or favorable trade winds along the route. Given how fast selection is known to act on migration patterns in other species (Berthold et al. 1992), one would expect that nonadaptive migration patterns would change over time so that members of the *swainsoni* lineage would begin taking the shortest route by migrating down the Pacific Coast.

Although the migration patterns of *ustulatus* and northwestern and eastern *swainsoni* populations are relatively well documented, the patterns of migration in Rocky Mountain breeding populations remain less clear. Current data are inadequate to determine whether Rocky Mountain and northwestern *swainsoni* populations expanded out of the same region at the end of the LGM, or even to map the current migratory pattern of Rocky Mountain populations. Available genetic data indicate that Rocky Mountain populations are indistinguishable from other *swainsoni* populations, thus suggesting that they share a similar evolutionary history. However, further research is needed to help clarify questions regarding the migration routes in Rocky Mountain populations of the *swainsoni* subspecies group.

## Breeding Habitat and Climate Zones

The *ustulatus* and *swainsoni* groups occupy different habitat types across their extensive breeding range, which has led some authors to suggest that ecological specialization may play a role in maintaining subspecies boundaries (Bent 1949). Populations of *ustulatus* are riparian woodland specialists during the breeding season (Grinnell and Miller 1944, Bent 1949, Verner and Boss 1980). Their connection to riparian habitat is so close that, in California, it has been suggested that the loss of this habitat type may restrict dispersal (Johnson and Geupel 1996). By contrast, eastern and northern *swainsoni* populations occupy mixed-hardwood and Boreal spruce–fir forests, respectively; and, as Bent (1949) pointed out, these forest types are distinct from those of their *ustulatus* relatives. However, western interior *swainsoni* populations are associated with both *ustulatus*-like riparian habitat and mixed-conifer forests in interior British Columbia (Campbell et al. 1997) and the Rocky Mountains (Evans Mack and Yong 2000). The overlap in habitat types occupied by coastal *ustulatus* and western interior *swainsoni* populations suggests that habitat specialization is most likely not an important isolating barrier and that differences in habitat use across the range of *swainsoni* populations reflects plasticity rather than specificity.

The most striking difference between the two groups is not in habitat specialization, but in the occupation of distinctive coastal and interior climate-vegetation zones that likely reflect their unique evolutionary histories. The shift between coastal and interior climates is correlated with hypothesized regions of secondary contact between numerous sister species and subspecies of birds that likely were isolated in separate refugia during the LGM (Johnson 1978, Cicero 2004). The crest of the Cascade and Coastal mountains and the Sierra Nevada impose a significant precipitation shadow that produces a continent-wide shift from moist, coastal coniferous forest to the cool, dry forests of interior western North America. A recent study of climatic space occupied by the *ustulatus* and *swainsoni* groups supports the idea that regions of secondary contact fall on the boundary between coastal and interior climatic zones (Ruegg et al. 2006a). Interior *swainsoni* populations occupy breeding habitat characterized by lower annual mean temperature, greater extremes in temperature, and less fluctuation in precipitation across the year compared with the coastal *ustulatus* group's breeding habitat (Ruegg et al. 2006a). The question that remains unanswered is whether adaptations to distinctive climatic regions play a role in limiting gene flow between the subspecies. For example, do the *ustulatus* and *swainsoni* groups differ in the timing of migration, molt, and reproduction, and do these differences influence mate choice and the fitness of hybrids? I revisit this topic below.

## Acoustic Divergence

Throughout its range, the Swainson's Thrush is recognizable by its upward-spiraling flutelike song. The flutelike quality of the advertising song results from a string of syllables that increase in frequency and decrease in amplitude towards the end of the vocalization (Dobson and Lemon 1977). Each individual has multiple song types with significant individual variation in syllable type and order (Dobson and Lemon 1977, Ruegg et al. 2006b; Fig. 4). Despite significant individual variation, recognizable song types are often consistent within populations (Ruegg et al. 2006b). The high degree of individual variation suggests a strong capacity for vocal learning, consistent with data from other oscines in which song learning has been demonstrated (reviewed in

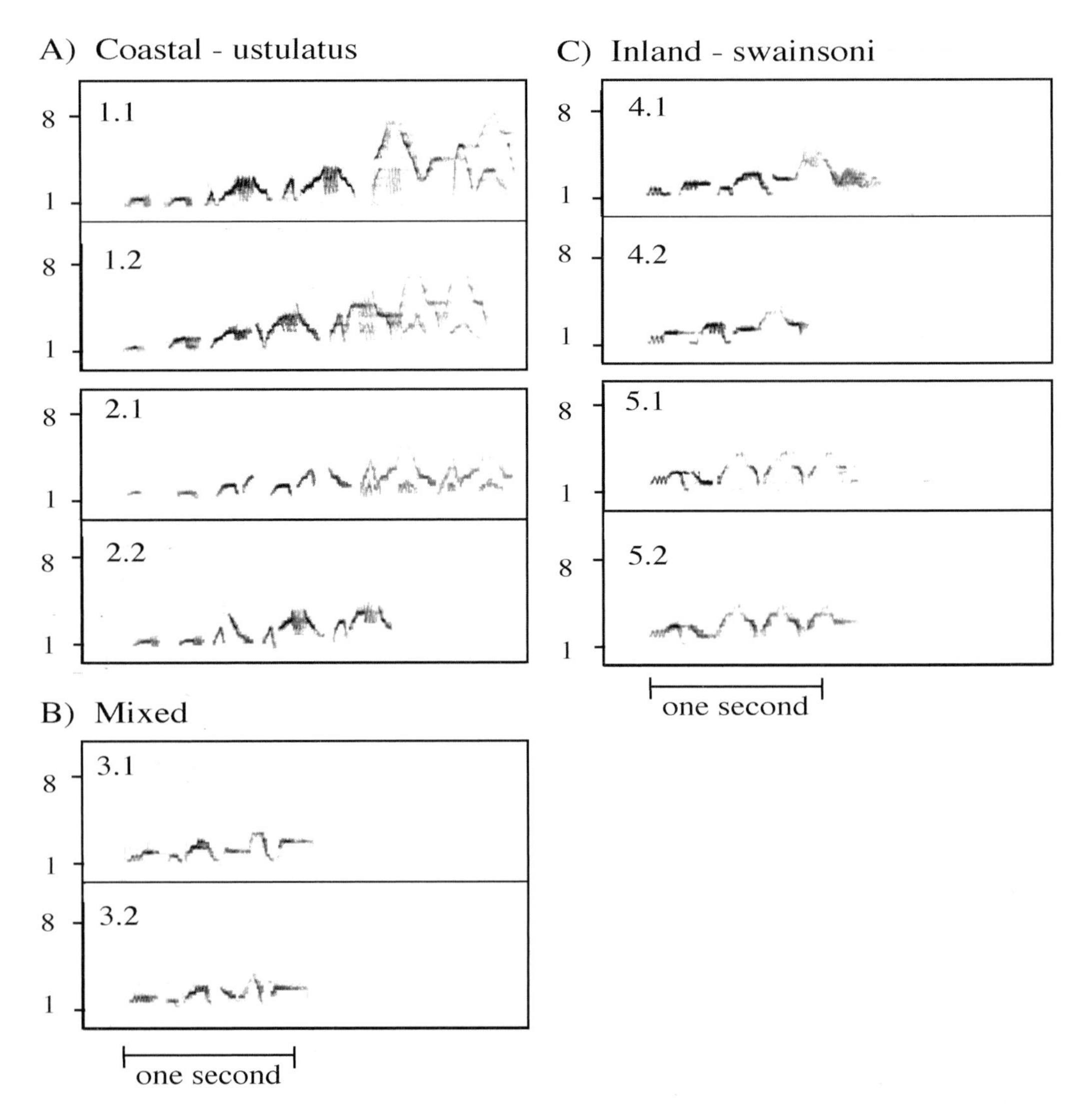


FIG. 4. Sonograms from two individuals from coastal *ustulatus* and inland *swainsoni* subspecies groups, as well as two individuals from a mixed population. The numbers in the upper left are the population number followed by the individual number. Although each individual has multiple song types, the figure illustrates the most common song type in the population (from Ruegg et al. 2006b).

Marler and Slabbekoorn 2004). Some geographic variation in the frequency of certain vocalizations, such as migration and nest-begging calls, has been reported (Evans Mack and Yong 2000), but a comprehensive survey of geographic variation in acoustic characteristics across the species' range is lacking. In addition, all work to date on the song of the Swainson's Thrush has focused on the advertising song (Dobson and Lemon 1977, Ruegg et al. 2006b). An especially useful future analysis would be to assess geographic variation in the frequency of call notes, which has been used previously to determine species status between Bicknell's and Gray-cheeked thrushes (Ouellet 1993).

In a study of genetic, ecological, and acoustic divergence in five populations of Swainson's Thrush—two from coastal temperate rainforest representing the *ustulatus* subspecies group, two from montane coniferous forest representing the *swainsoni* subspecies group, and one mixed population—Ruegg et al. (2006b) found a potential role for ecological selection in the evolution of vocal divergence. Overall, coastal songs were longer in duration and had lower initial frequencies than inland songs, whereas songs of the genetically more coastal, but ecologically more inland, mixed population were more similar to inland songs (Ruegg et al. 2006b). Ruegg et al. (2006b) hypothesized that the lower frequency and longer duration of coastal songs may result from the fact that sound transmission is more difficult in moist coastal forests, where background noise generated by rainfall is loud and mild winters allow for year-round growth of vegetation. Alternatively, shorter, higher-frequency inland songs may result from sound traveling more easily in drier mixed-coniferous forests, where understory vegetation is sparser. Further research is necessary to determine the extent to which habitat variables such as vegetation density and rainfall may influence sound transmission in the Swainson's Thrush.

The degree to which differences in song help maintain or promote further divergence between the *ustulatus* and *swainsoni* subspecies groups remains an open question and depends partly on the mechanism of song learning in this species. If song learning is restricted to the period before individuals disperse from their natal region, and if females prefer local song types, then song differences may present a barrier to gene flow (Ellers and Slabbekoorn 2003). However, if song learning occurs throughout life, individuals dispersing into new environments would have the capacity to adjust their song to the given ecological conditions and song would not present a barrier to gene flow (Ellers and Slabbekoorn 2003). Future research focused on the timing of song learning in the Swainson's Thrush will reveal the potential for acoustic variation to maintain or promote further divergence in this species.

### Hybrid Zone Dynamics

A detailed analysis of a hybrid zone between the *ustulatus* and *swainsoni* groups in the Coast Mountains of British Columbia revealed that the transition from *ustulatus* to *swainsoni* phenotypes and genotypes occurs over an 80-km region, which is concordant with the transition between coastal and interior climatic zones (Ruegg 2007). Using 15 amplified-fragment-length polymorphism (AFLP) markers and a Bayesian method for assigning individuals to hybrid classes (Anderson and Thompson 2002), Ruegg (2007) was able to distinguish between *ustulatus*, *swainsoni*, and hybrid individuals with >90% posterior probability. Morphological and plumage color analyses revealed that coastal *ustulatus* populations were larger, had relatively longer wings, and were more russet colored than inland *swainsoni* populations (Ruegg 2007). Clines in plumage color, body size, wing length, and nuclear genetic assignment (based on the AFLP markers) were coincident in position, whereas the mtDNA cline was shifted slightly southwest (Fig. 5). The width of the contact zone is approximately half the estimated dispersal distance of the Swainson's Thrush, which implies that there is a strong barrier to gene flow preventing the zone from becoming wider (Barton and Hewitt 1985, Barton and Gale 1993). The center of the hybrid zone was also characterized by low population density and the existence of pure *ustulatus* and *swainsoni* individuals as well as recent hybrids and backcrosses (Ruegg 2007). Ruegg (2007) evaluated a variety of models proposed to explain the maintenance of narrow hybrid zones in birds and concluded that the Swainson's Thrush hybrid zone had the characteristics of a tension zone in which dispersal into the zone was balanced by ecologically mediated selection against hybrids or premating barriers to gene flow, or both.

The correlation of the hybrid zone with the transition from coastal to interior climatic regions suggests that hybrid fitness and premating isolation in the Swainson's Thrush may be influenced by adaptations to distinctive climatic regions. For example, annual fluctuations in temperature and seasonality may result in differences in the timing of molt, migration, and reproduction. Similar scenarios have been proposed to help explain the maintenance of hybrid zones for the Baltimore Oriole (*Icterus galbula*) and Bullock's Oriole (*I. bullockii*) and for yellow-shafted and red-shafted Northern Flickers (*Colaptes auratus auratus* and *C. a. cafer*, respectively) (Moore and Price 1993). Rohwer and Manning (1990) hypothesized that differences between Baltimore and Bullock's orioles in timing of molt is an adaptation to mesic and xeric climates and that hybrids with intermediate timing of molt may be less fit. Alternatively, the situation in the Swainson's Thrush may be more similar to the situation in Black-capped Warblers (*Sylvia atricapilla*), in which populations

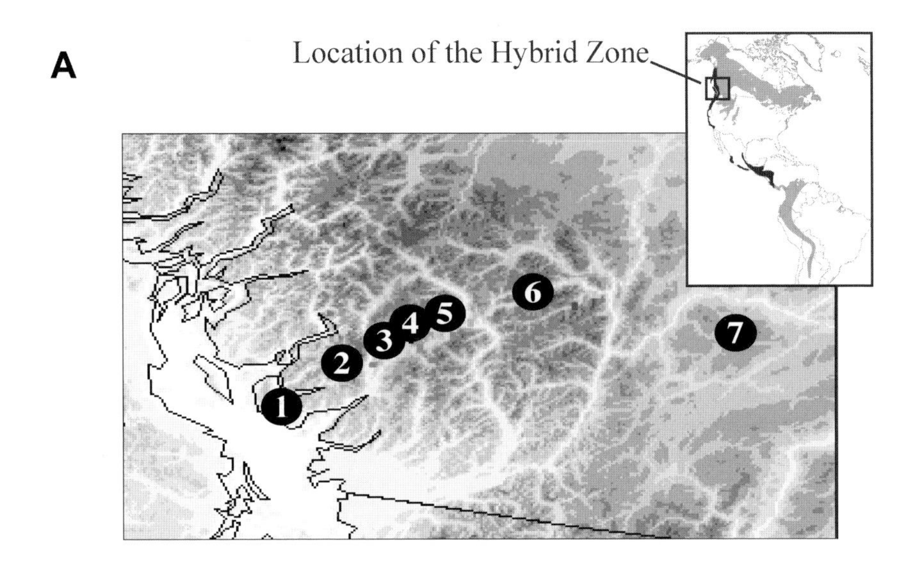


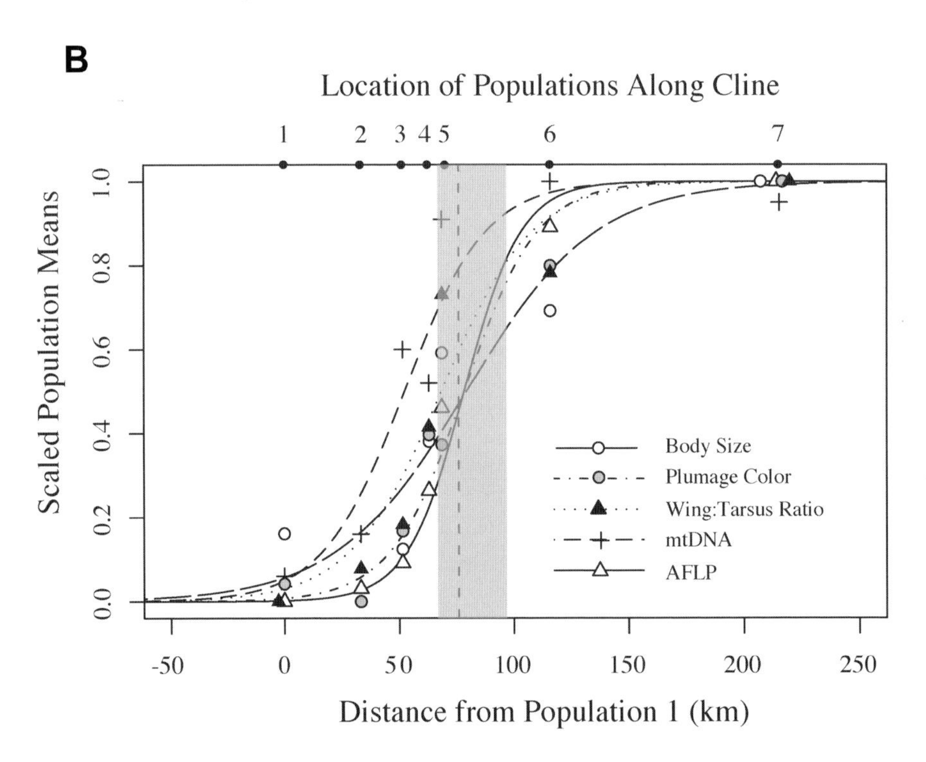


FIG. 5. Results from the hybrid-zone analysis. (A) Map of sampling locations in southwestern British Columbia. The smaller map in the upper right corner is the distribution of both forms, and the black rectangle represents the approximate region of the hybrid zone. (B) Shape and width of multiple character clines demonstrating concordant shifts from *ustulatus* to *swainsoni* phenotypes and genotypes across the hybrid zone.

with different wintering locations mate assortatively on the basis of arrival time (Bearhop et al. 2005). However, given evidence of hybridization between *ustulatus* and *swainsoni* populations in regions of overlap, premating isolation alone cannot explain the strong barrier to gene flow. To investigate the role of climate in hybrid-zone maintenance, future research should focus on how differences in the timing of migration, molt, and reproduction affect premating isolation and hybrid fitness.

### Implications for Speciation

From a speciation standpoint, the *ustulatus* and *swainsoni* subspecies groups reside in the interesting gray area between well-differentiated subspecies and recently diverged sister species. On one hand, the two groups share many characteristics of well-differentiated subspecies: (1) genetic divergence estimates between the groups is, in many cases, lower than levels of genetic divergence between many known sister taxa, but similar to levels of genetic divergence between subspecies; (2) the subspecies groups occupy distinctive climatic regions, but given the available data there is little to suggest that they are ecologically isolated; and (3) the two groups have differences in their advertising song, but the song differences are similar to dialect differences found within other species where song learning is prevalent. On the other hand, the two groups share many characteristics of recently diverged sister species: (1) *ustulatus* and *swainsoni* populations are diagnosable as distinct genetic entities using both mitochondrial and nuclear (AFLP) markers; (2) climatic models and genetic data both support the idea that the two groups have distinct evolutionary histories; and (3) cline analysis suggests that one hybrid zone is approximately half the estimated dispersal distance of the Swainson's Thrush, implying that there is a strong barrier to gene flow preventing the zone from becoming wider.

The AOU has traditionally relied on the biological species concept (BSC) of Dobzhansky (1937), which states that "species are systems of populations: the gene exchange between these systems is limited or prevented by a reproductive isolating mechanism or perhaps by a combination of several mechanisms." As discussed by Johnson et al. (1999), the gray area in the interpretation of the BSC for avian taxonomy is the extent to which hybridization is allowed before the two groups are considered subspecies rather than sister taxa. The situation in the Swainson's Thrush is comparable to that in Yellow-shafted and Red-shafted flickers (Moore and Price 1993), Myrtle and Audubon's warblers (*Dendroica coronata*; Barrowclough 1980), and Hermit and Townsend's warblers (*D. occidentalis* and *D. townsendi*; Rohwer and Wood 1998, Rohwer et al. 2001), in which sister groups are clearly diagnosable according to many criteria but are also known to hybridize extensively in restricted hybrid zones. Although the BSC allows for hybridization, a key component of Dobzhansky's (1937) definition is that there must be some evidence for reproductive isolation. In the case of the flickers and the Myrtle × Audubon's warblers, there is little evidence for reproductive isolation, and the two groups have been treated as one species. However, in the case of Hermit × Townsend's warblers, the high number of parentals in the center of the hybrid zone and the narrow width of the hybrid zone in relation to estimates of dispersal distance provide evidence for reproductive isolation and, as a result, the groups are considered separate species (AOU 1983). Similar to the situation in Hermit and Townsend's warblers, the Swainson's Thrush hybrid zone is narrow in relation to estimates of dispersal and contains a significant number of parentals within the centermost populations—thus indicating a strong potential for reproductive isolation between *ustulatus* and *swainsoni* individuals (Ruegg 2007). In addition, the Swainson's Thrush hybrid zone is correlated with the transition from coastal to interior climatic regions, which suggests a role for ecologically mediated reproductive isolation. However, additional research is necessary to identify the mechanisms of reproductive isolation that may be occurring to keep the hybrid zone from widening. If Ruegg's (2007) analysis is correct and climatic differences are important in limiting gene flow between the subspecies, then *ustulatus* and *swainsoni* populations will remain on distinct evolutionary trajectories as long as the coastal and interior climatic regions remain stable. Future research would benefit from analysis of additional hybrid-zone transects as well as tests for hybrid fitness and assortative mating within the hybrid zone.

Acknowledgments

I would like to thank Ned K. Johnson and Carla Cicero for introducing me to the study of ornithology. I also would like to thank J. V. Remsen, C. Cicero, and two anonymous reviewers for their comments on earlier drafts of this manuscript.

Literature Cited

Abbott, C. L., and M. C. Double. 2003. Genetic structure, conservation genetics, and evidence of speciation by range expansion in Shy and White-capped albatrosses. Molecular Ecology 12:2953–2963.

American Ornithologists' Union. 1983. Check-list of North American Birds, 6th ed. American Ornithologists' Union, Washington, D.C.

American Ornithologists' Union. 1998. Check-list of North American Birds, 7th ed. American Ornithologists' Union, Washington, D.C.

Anderson, E. C., and E. A. Thompson. 2002. A model-based method for identifying species hybrids using multilocus genetic data. Genetics 160:1217–1229.

Avise, J. C., and D. Walker. 1998. Pleistocene phylogeographic effects on avian populations and the speciation process. Proceedings of the Royal Society of London, Series B 265:457–463.

Barrowclough, G. F. 1980. Genetic and phenotypic differentiation in a wood warbler (genus *Dendroica*) hybrid zone. Auk 97:655–668.

Barton, N. H., and K. S. Gale. 1993. Genetic analysis of hybrid zones. Pages 13–45 *in* Hybrid Zones and the Evolutionary Process (R. G. Harrison, Ed.). Oxford University Press, New York.

Barton, N. H., and G. M. Hewitt. 1985. Analysis of hybrid zones. Annual Review of Ecology and Systematics 16:113–148.

Bearhop, S., W. Fiedler, R. W. Furness, S. C. Votier, S. Waldron, J. Newton, G. J. Bowen, P. Berthold, and K. Farnsworth. 2005. Assortative mating as a mechanism for rapid evolution of a migratory divide. Science 310: 502–504.

Bent, A. C. 1949. Life histories of North American thrushes, kinglets, and their allies. United States National Museum Bulletin 196.

Berthold, P., A. J. Helbig, G. Mohr, and U. Querner. 1992. Rapid microevolution of migratory behaviour in a wild bird species. Nature 360:668–670.

Bond, G. M. 1963. Geographic variation in the thrush *Hylocichla ustulata*. Proceedings of the United States National Museum 114:373–387.

Burg, T. M., A. J. Gaston, K. Winker, and V. L. Friesen. 2005. Rapid divergence and postglacial colonization in western North American Steller's Jays (*Cyanocitta stelleri*). Molecular Ecology 14:3745–3755.

Burg, T. M., A. J. Gaston, K. Winker, and V. L. Friesen. 2006. Effects of Pleistocene glaciations on population structure of North American Chestnut-backed Chickadees. Molecular Ecology 15:2409–2419.

Campbell, W. R., N. K. Dawe, I. McTaggart-Cowan, J. M. Cooper, G. W. Kaiser, M. C. E. McNall, and G. E. J. Smith. 1997. The Birds of British Columbia: Flycatchers through Vireos. University of British Columbia Press, Vancouver.

Cicero, C. 2004. Barriers to sympatry between avian sibling species (Paridae: *Baeolophus*) in tenuous secondary contact. Evolution 58: 1573–1587.

Crochet, P. A., J. Z. Chen, J. M. Pons, J. D. Lebreton, P. D. N. Herbert, and F. Bonhomme. 2003. Genetic differentiation at nuclear and mitochondrial loci among large white-headed gulls: Sex-biased interspecific gene flow? Evolution 57:2865–2878.

Dobson, C. W., and R. E. Lemon. 1977. Markovian versus rhomboidal patterning in the song of Swainson's Thrush. Behaviour 62:3–4.

Dobzhansky, T. 1937. Genetics and the Origin of Species. Columbia University Press, New York.

Ellers, J., and H. Slabbekoorn. 2003. Song divergence and male dispersal among bird populations: A spatially explicit model testing the role of vocal learning. Animal Behaviour 65: 671–681.

Evans Mack, D., and W. Yong. 2000. Swainson's Thrush (*Catharus ustulatus*). *In* The Birds of North America, no. 540 (A. Poole and F. Gill, Eds.). Birds of North America, Philadelphia.

Gibbs, H. L., R. J. G. Dawson, and K. A. Hobson. 2000. Limited differentiation in microsatellite DNA variation among northern populations of the Yellow Warbler: Evidence for male-biased gene flow? Molecular Ecology 9:2137–2147.

Grinnell, J., and A. H. Miller. 1944. The distribution of the birds of California. Pacific Coast Avifauna, no. 27.

Johnson, M. D., and G. R. Geupel. 1996. The importance of productivity to the dynamics of a Swainson's Thrush population. Condor 98:133–141.

Johnson, N. K. 1978. Patterns of avian geography and speciation in the Intermountain Region. Great Basin Naturalist Memoirs 2:137–159.

Johnson, N. K., and C. Cicero. 2004. New mitochondrial DNA data affirm the importance of Pleistocene speciation in North American birds. Evolution 58:1122–1130.

Johnson, N. K., J. V. Remsen, Jr., and C. Cicero. 1999. Resolution of the debate over species concepts in ornithology: A new comprehensive biologic species concept. Pages 1470–1482 *in* Acta XXII Congressus Internationalis Ornithologici (N. J. Adams and R. H. Slotow, Eds.). BirdLife South Africa, Johannesburg.

Klicka, J., R. M. Zink, J. C. Barlow, W. B. McGillivray, and T. J. Doyle. 1999. Evidence supporting the recent origin and species status of the Timberline Sparrow. Condor 101:577–588.

Lovette, I. J. 2005. Glacial cycles and the tempo of avian speciation. Trends in Ecology and Evolution 20:57–59.

Marler, P., and H. Slabbekoorn, Eds. 2004. Nature's Music: The Science of Birdsong. Elsevier Academic Press, San Francisco.

Marshall, H. D., and A. J. Baker. 1998. Rates and patterns of mitochondrial DNA sequence evolution in fringilline finches (*Fringilla* spp.) and the Greenfinch (*Carduelis chloris*). Molecular Biology and Evolution 15:638–646.

Moore, W. S., and J. T. Price. 1993. Nature of selection in the Northern Flicker hybrid zone and its implications for speciation theory. Pages 196–225 *in* Hybrid Zones and the Evolutionary Process (H. Richard, Ed.). Oxford University Press, New York.

Ouellet, H. 1993. Bicknell's Thrush: Taxonomic status and distribution. Wilson Bulletin 105: 545–572.

Outlaw, D. C., G. Voelker, B. Mila, and D. J. Girman. 2003. Evolution of long-distance migration in and the historical biogeography of *Catharus* thrushes: A molecular phylogenetic approach. Auk 120:299–310.

Phillips, A. R. 1991. The Known Birds of North and Middle America, part II. Published by the author. Denver, Colorado.

Pielou, E. C. 1991. After the Ice Age: The Return of Life to Glaciated North America. University of Chicago Press, Chicago, Illinois.

Questiau, S., M. C. Eybert, A. R. Gaginskaya, L. Gielly, and P. Taberlet. 1998. Recent divergence between two morphologically differentiated subspecies of Bluethroat (Aves: Muscicapidae: *Luscinia svecica*) inferred from mitochondrial DNA sequence variation. Molecular Ecology 7:239–245.

Ramos, M. A., and D. W. Warner. 1980. Analysis of North American subspecies of migrant birds wintering in Los Tuxtlas, southern Veracruz, Mexico. Pages 173–180 *in* Migrant Birds in the Neotropics: Ecology, Behavior, Distribution, and Conservation (A. Keast and E. S. Morton, Eds.). Smithsonian Institution Press, Washington, D.C.

Rohwer, S., E. Bermingham, and C. Wood. 2001. Plumage and mitochondrial DNA haplotype variation across a moving hybrid zone. Evolution 55:405–422.

Rohwer, S., and J. Manning. 1990. Differences in timing and number of molts for Baltimore and Bullock's orioles: Implications to hybrid fitness and theories of delayed plumage maturation. Condor 92:125–140.

Rohwer, S., and C. Wood. 1998. Three hybrid zones between Hermit and Townsend's warblers in Washington and Oregon. Auk 115:284–310.

Ruegg, K. C. 2007. The origin and maintenance of a migratory divide in the Swainson's Thrush (*Catharus ustulatus*) and its implications for speciation. Ph.D. dissertation, University of California, Berkeley.

Ruegg, K. C., R. J. Hijmans, and C. Moritz. 2006a. Climate change and the origin of migratory pathways in the Swainson's Thrush (*Catharus ustulatus*). Journal of Biogeography 33: 1172–1182.

Ruegg, K. C., H. Slabbekoorn, S. M. Clegg and T. B. Smith. 2006b. Divergence in mating signals correlates with ecological variation in a migratory songbird, the Swainson's Thrush (*Catharus ustulatus*). Molecular Ecology 15: 3147–3156.

Ruegg, K. C., and T. B. Smith. 2002. Not as the crow flies: A historical explanation for circuitous migration in Swainson's Thrush (*Catharus ustulatus*). Proceedings of the Royal Society of London, Series B 269:1375–1381.

Verner, J., and A. Boss. 1980. California wildlife and their habitats: Western Sierra Nevada. U.S. Department of Agriculture, Forest Service General Technical Report PSW-37.

Winker, K., and C. L. Pruett. 2006. Seasonal migration, speciation, and morphological convergence in the genus *Catharus* (Turdidae). Auk 123:1052–1068.

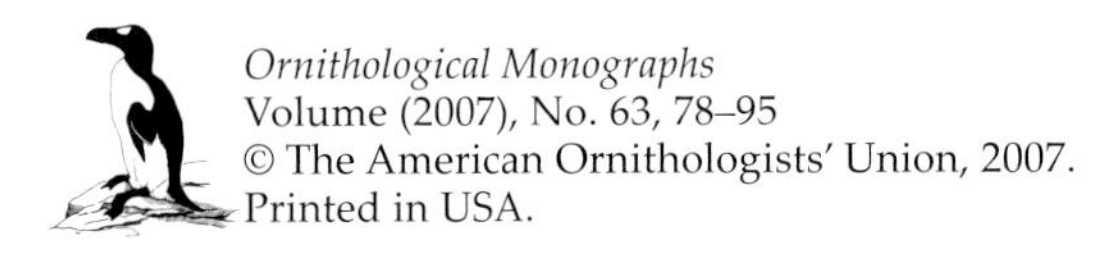
*Ornithological Monographs*
Volume (2007), No. 63, 78–95


CHAPTER 7

# NARROW CONTACT OF DESERT SAGE SPARROWS (*AMPHISPIZA BELLI NEVADENSIS* AND *A. B. CANESCENS*) IN OWENS VALLEY, EASTERN CALIFORNIA: EVIDENCE FROM MITOCHONDRIAL DNA, MORPHOLOGY, AND GIS-BASED NICHE MODELS

Carla Cicero[1] and Ned K. Johnson[2]

*Museum of Vertebrate Zoology and Department of Integrative Biology, University of California, Berkeley, California 94720, USA*

Abstract.—We investigated the distribution of interior subspecies of Sage Sparrows (*Amphispiza belli nevadensis* and *A. b. canescens*) in Owens Valley, eastern California. Our primary goals were to establish the geographic limits of subspecies' nesting ranges in this region, to examine whether the two forms are in contact, and to elucidate their evolutionary history. Mitochondrial DNA (mtDNA) analysis and discriminant-function classification of size pointed to a narrow contact zone near Bishop, Inyo County, with evidence for limited sympatry and possible intergradation or hybridization. Estimates of mtDNA gene flow are low. The contact zone occurs toward the northern end of a bioclimatic gradient that spans the valley from the Great Basin in the north (cold and wet) to the Mojave Desert in the south (hot and dry). Ecological and bioclimatic data show slightly different patterns but generally suggest a transition that coincides with the distributional limits of the two forms. According to population genetic measures, both subspecies have expanded rapidly into Owens Valley. Because *A. b. nevadensis* and *A. b. canescens* occupy distinctive vegetation–climate associations, local adaptation to conditions in this region determines the extent to which the two forms can coexist at the margins of their respective ranges. *Received 30 July 2006, accepted 9 February 2007.*

Resumen.—En este trabajo, mostramos los resultados de una intensa investigación sobre la distribución de las subespecie de interior del Zacatonero de artemisa (*Amphispiza belli nevadensis* and *A. b. canescens*) en el valle de Owens, al este de California. El objetivo principal de nuestro estudio fue establecer los límites geográficos de los rangos de nidificación de las subespecies en la región, con el fin de determinar si ambas formas se encuentran en contacto, y así elucidar su historia evolutiva. Los análisis de ADN mitocondrial y una función discriminante para el tamaño de las aves revelaron una delgada zona de contacto cercana a Bishop, condado de Inyo, con evidencia de simpatría limitada y posible intergradación o hibridación. Las estimaciones del flujo génico para el mtDNA dan valores bajos. La zona de contacto ocurre en el límite norte del gradiente bioclimático que cubre el valle desde la gran cuenca hidrográfica en el norte (frío/ húmedo) hasta el desierto de Mojave en el sur (caluroso/seco). Los datos ecológicos y bioclimáticos muestran patrones levemente distintos pero, por lo general, sugieren una transición que coincide con los límites de distribución de las dos formas. De acuerdo con los parámetros poblacionales estudiados, las dos subespecies se han extendido rápidamente en el valle de Owens. Dado que *A. b. nevadensis* y *A. b. canescens* ocupan asociaciones diferentes de vegetación-clima, las adaptaciones locales a las condiciones de cada región determinan hasta qué punto las dos formas pueden coexistir en los márgenes de sus rangos respectivos.

[1]E-mail: ccicero@berkeley.edu
[2]Deceased 11 June 2003.

THIS IS ONE of a series of papers aimed at understanding the systematic relationships of Sage Sparrow (*Amphispiza belli*) subspecies in western North America. Of the five taxa currently recognized (*A. b. belli*, *A. b. clementeae*, *A. b. cinerea*, *A. b. canescens*, and *A. b. nevadensis*; Martin and Carlson 1998), three (*A. b. belli*, *A. b. canescens*, and *A. b. nevadensis*; Fig. 1) have been the focus of ongoing investigations that began in 1977 with the collection of samples of specimens by N.K.J. for a study of morphometric and allozyme variation (Johnson and Marten 1992). In addition to the allozyme study, Johnson and Cicero (1991) used mitochondrial DNA (mtDNA) sequencing of a small fragment of the cytochrome-*b* gene to assess genetic relationships among the three subspecies. Over the past several years, we have extended this analysis to include significantly larger numbers of samples using both polymerase chain reaction (PCR) and restriction fragment length polymorphism (RFLP) and complete sequencing of cytochrome *b*, and to correlate genetic and phenotypic patterns with bioclimatic gradients using geographic information system (GIS) modeling (C. Cicero and N. K. Johnson unpubl. data). Most recently, Cicero and Johnson (2006) analyzed morphological variation between *A. b. canescens* and *A. b. nevadensis* to refute the claim (Patten and Unitt 2002) that these two forms are not diagnosable and should be synonymized. Finally, work in progress entails analysis of vocal variation in relation to morphological and molecular differences where *A. b. canescens* and *A. b. nevadensis* approach in Owens Valley, eastern California, (Mono and Inyo counties). This intensive investigation of potential contact in Owens Valley was the focus of one of N.K.J.'s final field trips in 2002.

Owens Valley stretches north–south for ~160 km and is bounded by the Sierra Nevada to the west and the White and Inyo mountains to the east. The mountains include the highest peak in the contiguous United States (Mount Whitney, at >4,400 m), and the valley floor is at 1,200 m, which makes it one of the deepest valleys in the United States. Owens Valley is also the westernmost "graben" in the Basin and Range Province, a downdropped block of land between two vertical faults. Desert shrub communities typical of the rain shadow of the Sierra Nevada dominate the vegetation of the valley floor. These plant associations grade sharply from the Great Basin in the north to the Mojave Desert in the south. The Owens River runs through the valley, providing important riparian and wetland habitat that has suffered significant losses because of water use over the past century.

Grinnell and Miller (1944) reported that the northern limit of *A. b. canescens*, which in California is a summer resident of desert scrub in the northern and western Mojave Desert and in the San Joaquin Valley, occurs in the vicinity of Benton in Benton Valley, Mono County. They proposed that it intergrades there with *A. b. nevadensis*, a significantly larger subspecies (Johnson and Marten 1992, Cicero and Johnson 2006) that breeds primarily in sagebrush (*Artemesia tridentata*) of the Great Basin region. On the basis of allozymes and morphology, however, Johnson and Marten (1992) firmly concluded that populations near the north end of the White Mountains (including Benton Valley as well as Chalfant Valley just to the south) are *A. b. nevadensis*. Although it is possible that the area of contact or intergradation has moved southward during the past 60 years, there is no evidence to support this hypothesis. Furthermore, if any change had occurred, one would expect a northward shift of the warmer-adapted *A. b. canescens* associated with increased summer temperatures (e.g., Johnson 1994).

From their findings, Johnson and Marten (1992:18) surmised that

> if a region of contact and intergradation exists between these forms it must be situated somewhere in the 100 mile [160 km] corridor between the southernmost certain breeding population of *A. b. nevadensis* (Chalfant Valley) and the northernmost definite nesting population of *A. b. canescens* (Coso Junction [see Fig. 1]). Investigation of this zone should yield information on the possible biologic species status of these two strongly differentiated forms.

To follow up on this earlier work, we used molecular, morphological, and bioclimatic data to examine distributional, evolutionary, and ecological patterns within *A. belli* in eastern California. Specifically, our goals were to (1) establish the nesting limits of *A. b. canescens* and *A. b. nevadensis* in or near the Owens Valley; (2) examine whether the two forms may come into contact and, if so, delineate the contact zone;

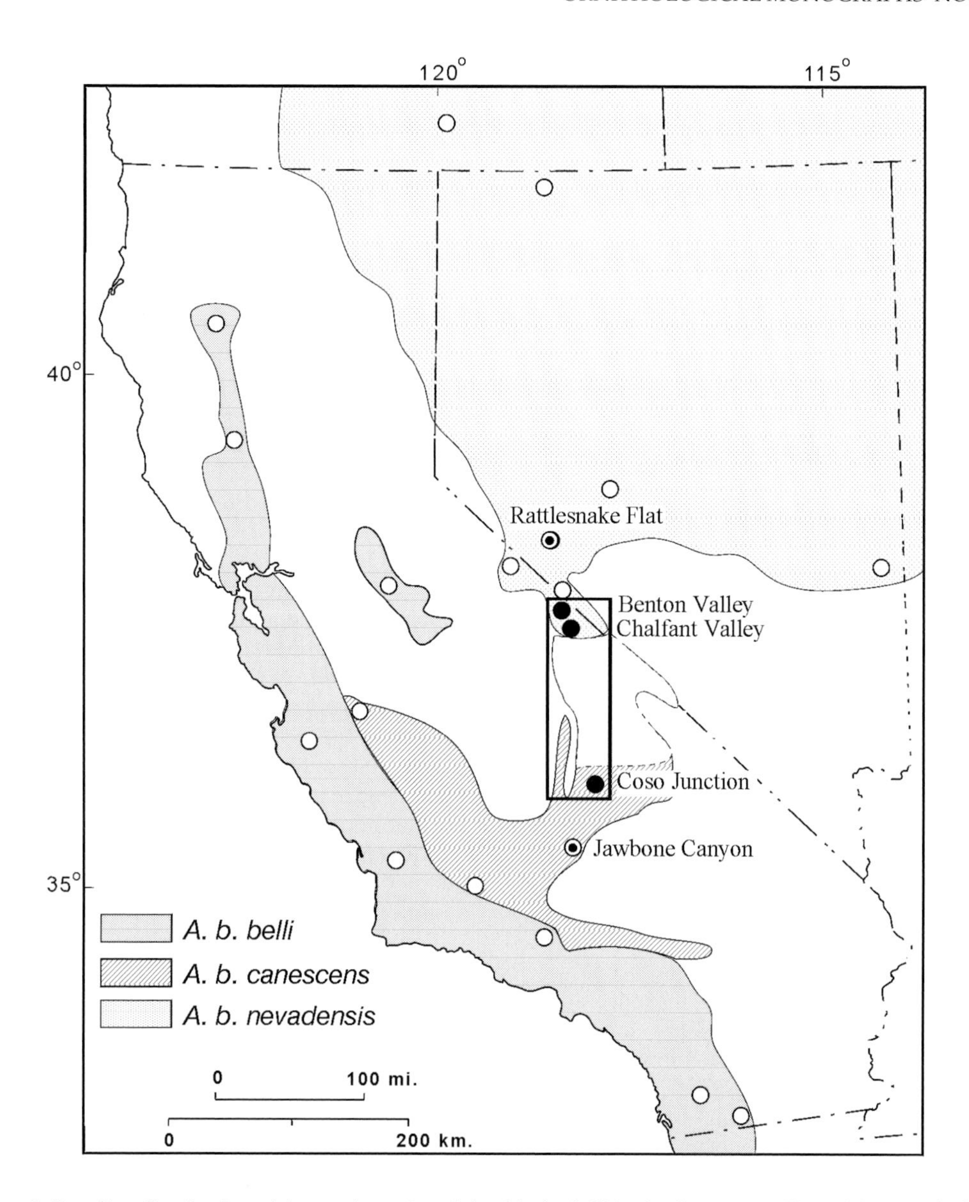


FIG. 1. Breeding distribution of three subspecies of *Amphispiza belli* in the far western United States. The box outlines the study area of potential contact between *A. b. nevadensis* and *A. b. canescens* in Owens Valley, eastern California. Circles show 21 of 22 samples analyzed previously by Johnson and Marten (1992); one sample is not shown because it contained postbreeding, upslope migrants. Open circles have been analyzed elsewhere for mtDNA (C. Cicero and N. K. Johnson unpubl. data) and were not included in the present study. Two samples (Rattlesnake Flat and Jawbone Canyon) were included as references outside the study area. Three samples (Benton Valley, Chalfant Valley, and Coso Junction) show definite nesting limits of the two forms before this study. Grinnell and Miller (1944) reported that specimens near Benton represent *A. b. canescens* or intergrade toward *A. b. nevadensis*, whereas Johnson and Marten (1992) concluded that populations at Benton and Chalfant valleys are definitely *A. b. nevadensis* (see text).

and (3) understand the evolutionary history of the species in this region. The results of the present study, together with additional mtDNA and nuclear analyses and a study of vocal variation in the Owens Valley (C. Cicero and N. K. Johnson unpubl. data), will confirm whether the two forms should be regarded as biological species. The taxonomic relationship of *A. b. nevadensis* and *A. b. canescens* to other subspecies, especially *A. b. belli*, will be discussed in a separate paper (C. Cicero and N. K. Johnson unpubl. data).

## Materials and Methods

*Field work and specimens examined.*—N.K.J. began collecting specimens in the Owens Valley for this study in spring 1991, and subsequent field seasons often included a trip to the region for additional sampling (and trout fishing whenever possible). The last trip was taken by C.C. in 2004. During this period, we collected a total of 198 specimens with tissues, which supplemented existing samples ($n$ = 357) taken for the broader analysis of allozymes and morphology (Johnson and Marten 1992). Specimens were prepared in the field as study skins, and tissues (heart, liver, muscle, kidney) were preserved in liquid nitrogen for later storage in ultra-low (–80°C) freezers. In general, our efforts focused on recording songs from breeding males and then targeting those same individuals for collection. Although detailed study of vocal variation will be presented elsewhere, the availability of vocal, morphological, and genetic data from the same birds provides a powerful data set for assessing biological species status. All specimens, tissues, and recordings are deposited in the collections of the Museum of Vertebrate Zoology, University of California, Berkeley (see Acknowledgments).

Samples from Owens Valley were grouped into 14 geographic areas for analysis, with two additional samples (Rattlesnake Flat, Jawbone Canyon) included as references from outside the area of potential contact (see Table 1, Fig. 2, and Appendix). Five of the 16 samples were studied previously by Johnson and Marten (1992), with supplemental collections at two locations (Benton Valley, Chalfant Valley) in 2002 and 2004. As noted above, Benton and Chalfant valleys represent the southernmost known nesting limit of *A. b. nevadensis*, and Coso Junction is the northernmost limit of known nesting for *A. b. canescens* (Johnson and Marten 1992).

*Molecular analyses.*—A broader geographic study (C. Cicero and N. K. Johnson unpubl. data) of cytochrome-*b* variation in 304 individuals of *A. b. belli, A. b. canescens,* and *A. b. nevadensis* (17 of 22 sample areas analyzed by

Table 1. Samples of *Amphispiza belli* analyzed for mtDNA ($n$) and morphology (adult males only) in and near Owens Valley. Distribution of distinct cytochrome-*b* haplotype groups (AC, AD, BC) is also indicated. See Figures 1 and 2 for sample locations.

| | Sample name and number [a] | $n$ [b] | Number of males [c] | AC | AD | BC |
|---|---|---|---|---|---|---|
| Ref | Rattlesnake Flat* | 15 (2) | 15 | | | 15 |
| 1 | Benton Valley* | 31 (4) | 25 | | 1 | 30 |
| 2 | Chalfant Valley* | 23 (2) | 16 | | | 23 |
| 3 | Volcanic Tableland | 18 (2) | 16 | | | 18 |
| 4 | East of Laws | 14 (3) | 13 | | 2 | 12 |
| 5 | West of Laws | 13 (4) | 13 | | 2 | 11 |
| 6 | Tungsten Hills | 15 (4) | 15 | | 9 | 6 |
| 7 | Horton Creek | 16 (3) | 16 | | 15 | 1 |
| 8 | Southwest of Bishop | 16 (4) | 15 | | 13 | 3 |
| 9 | South of Bishop | 18 (6) | 17 | 2 | 12 | 4 |
| 10 | Southeast of Bishop | 13 (2) | 13 | | 11 | 2 |
| 11 | West of Black Canyon | 13 (4) | 13 | | 9 | 4 |
| 12 | Big Pine | 15 (2) | 14 | | 14 | 1 |
| 13 | Independence | 20 (2) | 20 | | 19 | 1 |
| 14 | Coso Junction* | 18 (6) | 8 | 2 | 14 | 2 |
| Ref | Jawbone Canyon* | 15 (3) | 11 | 1 | 14 | |
| Totals | | 273 (53) | 240 (238) | 5 | 135 | 133 |

[a] Asterisks indicate samples analyzed previously by Johnson and Marten (1992). Two sites (Rattlesnake Flat and Jawbone Canyon) were included as references away from the Owens Valley area of potential contact. Haplotype group designation follows C. Cicero and N. K. Johnson (unpubl. data). Group AC, which is rare in *A. b. canescens* from Owens Valley and the Mojave Desert, is more common in populations of that taxon from the San Joaquin Valley of California.

[b] Number of individuals analyzed with PCR–RFLP (number of individuals sequenced in parentheses).

[c] Total number in parentheses represents specimens that were not missing morphometric data.

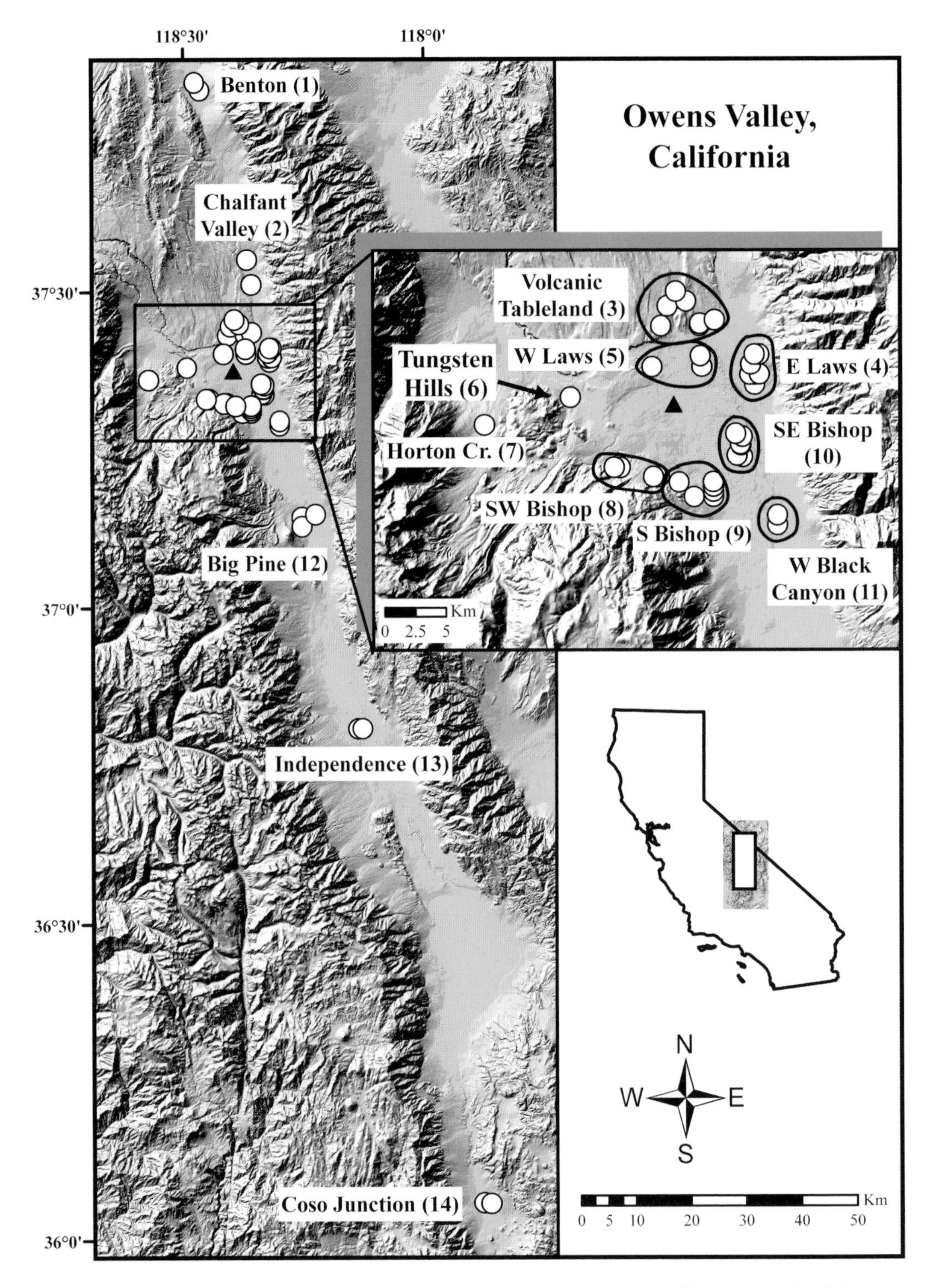


Fig. 2. Sampling locations for specimens of *Amphispiza belli* from the Owens Valley transect. Localities were grouped into 14 sample areas for analysis (see Table 1). Black triangle indicates location of Bishop, Inyo County, California. Bottom scale (inset) refers to the relief map of the entire valley.

Johnson and Marten 1992; see Fig. 1) revealed three primary haplotype groups, two of which are dominant in *A. b. nevadensis* and *A. b. canescens*, respectively, as they approach Owens Valley. These data were based on a combination of DNA sequencing and PCR–RFLP, a useful technique for classifying large numbers of individuals with respect to haplotype in phylogeographic or hybrid-zone studies (e.g., Rohwer et al. 2001, Ruegg and Smith 2002, Cicero 2004). We used the same approach to identify individuals of *A. belli* collected at each of the sample areas from Rattlesnake Flat to Jawbone Canyon ($n = 273$; Table 1).

Whole genomic DNA was extracted from fresh frozen tissue using a DNeasy tissue kit (Qiagen, Valencia, California), and a ~900 base-pair (bp) fragment of cytochrome *b* was amplified using primers L14987 and H15916 (Cicero and Johnson 2001; Table 2). A negative control was included in all extraction and amplification experiments. Two diagnostic restriction enzymes (Alu I, Hinf I) were used to digest the PCR products using the same protocol described by Cicero (2004), and fragments were visualized on 2.5% Nusieve:0.5% Seakem agarose gels stained with ethidium bromide. Individuals then were scored according to haplotype pattern. To confirm haplotype designations, and for population genetic analyses, a subset of 53 individuals (Table 1) was amplified and sequenced for the entire cytochrome-*b* gene (primers L14851–H15304, L15236–H15916, L15557–H16065 in Cicero and Johnson 2001; Table 2). Sequences were obtained with an ABI Prism 377 automated sequencer (Applied Biosystems, Foster City, California).

Sequences were analyzed for several population genetic measures using ARLEQUIN, version 3.01 (Excoffier et al. 2005) and DNASP, version 4.10 (Rozas et al. 2003). Analyses included (1) nucleotide diversity and its variance (Nei 1987), which estimates the mean number of nucleotide differences among haplotypes; (2) hierarchical analysis of molecular variance (AMOVA), which estimates genetic diversity among versus within subspecies; (3) Fu's $F_S$ statistic (Fu 1997), which—assuming neutrality—can be used to infer demographic population expansion; (4) mismatch distributions (Slatkin and Hudson 1991, Rogers and Harpending 1992), which compare the frequency of observed versus expected pairwise differences between haplotypes to see whether they show patterns of an expanding population; and (5) pairwise *M*, which uses the island model of population structure (Wright 1951) to estimate gene flow (*Nm*) by the relationship $M = (1 - F_{ST})/2 F_{ST}$.

*Morphometrics.*—Seven linear external measurements (mm) were taken on museum skins according to standard protocols (Johnson 1980, Cicero 1996) with dial calipers: wing length; tail length; bill length, depth, and width; tarsus length; and middle toe length without claw. In addition, body mass was recorded in the field and transformed to cube root for analysis. Morphological data were recorded from the same specimens used in molecular analyses to assess individual concordance between phenotype and genotype. Because our field sampling emphasized singing birds to obtain tape-recorded songs, most of the specimens were males; thus, morphological data were limited to adult males ($n = 240$; Table 1), whereas the molecular data set ($n = 273$) also included adult females. In addition to singing, all males were in breeding condition on the basis of enlarged gonads.

TABLE 2. Population genetic measures for two subspecies of *Amphispiza belli*, subdivided by geographic region.[a] Values are based on complete sequences (1,143 bp) of the cytochrome-*b* gene. Percentage of variation among subspecies within regions is indicated in parentheses (AMOVA).

| Group | Sample size | Number of unique haplotypes | Nucleotide diversity | Fu's $F_S$ [b] |
|---|---|---|---|---|
| **Owens Valley (82.5–95.3%)** | | | | |
| *nevadensis* | 15 | 4 | 0.0004 ± 0.0004 | −2.369* |
| *canescens* [c] | 21 | 9 | 0.0042 ± 0.0024 | −0.074 |
| | 18 | 6 | 0.0011 ± 0.0008 | −1.908 |
| **Great Basin and Mojave Desert–San Joaquin Valley (57.5%)** | | | | |
| *nevadensis* | 15 | 4 | 0.0004 ± 0.0004 | −2.369* |
| *canescens* | 16 | 7 | 0.0078 ± 0.0043 | 2.963 |

[a] Analysis is based on sequenced specimens identified either as "pure" *A. b. nevadensis* or "pure" *A. b. canescens* on the basis of both diagnostic mtDNA haplotypes and discriminant-function scores (*nevadensis*, haplotype group BC, score > 0; *canescens*, haplotype groups AD or AC, score < 0; see Figs. 3 and 4). Thus, possible hybrid or intermediate individuals within Owens Valley (i.e., those with a haplotype diagnostic of one subspecies but morphologically classified as the other subspecies) were excluded from analysis.

[b] Values significant at $P \leq 0.05$ are indicated by an asterisk. Fu's $F_S$ for *A. b. canescens* from Owens Valley (−1.908) is barely nonsignificant ($P = 0.07$).

[c] Upper values for *A. b. canescens* include all "pure" individuals (see footnote a) in haplotype groups AD or AC (AMOVA among-subspecies variation = 82.5%). Lower values exclude three birds in the haplotype group AC (two at Coso Junction, one at Jawbone Canyon; Table 1), which characterizes *A. b. belli* and San Joaquin populations of *A. b. canescens* (C. Cicero and N. K. Johnson unpubl. data) but is rare in the Owens Valley region (AMOVA among-subspecies variation = 95.3%).

[d] Patterns outside the Owens Valley region are shown for comparison. The lower among-subspecies variation (AMOVA = 57.5%) is attributable to the presence of two distinct haplotype groups within *A. b. canescens* (C. Cicero and N. K. Johnson unpubl. data). Rattlesnake Flat, Benton Valley, Chalfant Valley, Coso Junction, and Jawbone Canyon were included in both analyses (see text).

Definitive breeding status is an important criterion when including individuals for geographic studies (Cicero and Johnson 2006).

Morphological variation was studied by stepwise discriminant-function analysis (DFA) using STATISTICA, version 6.0 (Statsoft, Tulsa, Oklahoma). Specimens were grouped by geographic sample for analysis, using Rattlesnake Flat and Jawbone Canyon as references away from Owens Valley. In addition, specimens were classified according to haplotype group (AD, BC; Table 1) to assess morphological versus genetic patterns at the level of individual birds. The five individuals who shared a third rare haplotype in the region (AC; Table 1) were combined with haplotype AD, because both represent *A. b. canescens* on a broader geographic scale (C. Cicero and N. K. Johnson unpubl. data). Comparison of AC and AD individuals in this region did not reveal any significant differences in size.

*Bioclimatic and landcover data.*—Bioclimatic and landcover data were obtained from online sources to examine climatic and vegetational trends across the Owens Valley region (Benton Valley to Coso Junction). For the bioclimatic analysis, 11 temperature (°C) and 8 precipitation (mm) variables were downloaded from the WORLDCLIM data set (see Acknowledgments). Unique locations ($n = 40$) for the 14 core sample areas were georeferenced to obtain latitude and longitude coordinates in decimal degrees, and bioclimatic variables were extracted for points at each site using DIVA-GIS, version 5.2 (Hijmans et al. 2005). Variables were analyzed across sites with principal-component analysis (STATISTICA, version 6.0). In addition, ecological-niche models for *A. b. nevadensis* and *A. b. canescens* were generated, along with those for *A. b. belli*, by the maximum-entropy method (Maxent; Elith et al. 2006; Phillips et al. 2004, 2006), using BIOCLIM variables and range-wide georeferenced localities for breeding specimens (C. Cicero and N. K. Johnson unpubl. data); these were used to visualize predicted distributions of the two subspecies in the Owens Valley.

Landcover data, including 77 landcover classes in the wildlife habitat relationships (WHR) system and a hierarchical classification of WHR types into 10 "major landcover" classes (WHR10), were obtained from the California Department of Forestry Fire and Resource Assessment Program (see Acknowledgments). Major landcover classes were mapped onto the Owens Valley region to examine geographic concordance between habitat type, Maxent predictions, and molecular or morphological transitions. All maps were generated using ARCMAP, version 9.1 (ESRI, Redlands, California).

## Results

*Sharp transition of cytochrome-*b *haplotypes in the northern Owens Valley.*—Of the 273 samples analyzed genetically using PCR–RFLP, including Rattlesnake Flat and Jawbone Canyon, 48.7% ($n = 133$) were classified in the BC haplotype group and 49.5% ($n = 135$) in the AD group (Table 1); the remaining five individuals (1.8%) had a haplotype characteristic of *A. belli* populations farther to the west (C. Cicero and N. K. Johnson unpubl. data). On a broader geographic scale, BC is typical of *A. b. nevadensis*, whereas AD is most common in *A. b. canescens* from the Mojave Desert. The distribution of haplotype groups across the 14 core sample areas in the Owens Valley region was not random. Thus, BC was dominant in samples from Benton Valley, Chalfant Valley, Volcanic Tableland, and near Laws, an old railroad station northeast of Bishop, Inyo County (Table 1 and Fig. 2); all these sites occur at the extreme northern end of the Owens Valley or just to the north. South of Bishop, the dominant haplotype group changed abruptly from BC to AD, a pattern that continued southward to Coso Junction at the extreme southern end of the Owens Valley region. Haplotype AD also dominated at Horton Creek, a tributary of the Owens River that flows north-northeast ~16 km west of Bishop. Interestingly, this haplotype appears to be rare in the more northern samples (only 5.1% of individuals from Benton Valley to Laws), whereas BC occurs in low to rare frequency all the way from Bishop to Coso Junction (15.3% of individuals). The most mixed sample occurs to the west of Bishop at Tungsten Hills, where both haplotypes were found in high frequency (AD = 60%; BC = 40%). Overall, estimated mtDNA gene flow between the two subspecies in this region was low ($M = 0.024$–$0.106$, depending on whether the rare AC haplotype group was included in the analysis); values greater than one are expected to prevent populations from differentiating solely because of drift (Slatkin 1993, Mills and Allendorf 1996).

*Morphological patterns are congruent with genetic data.*—Patterns of size variation detected by DFA generally were congruent with cytochrome-*b* variation in haplotype groups. Plots of discriminant-function scores for the two reference samples away from Owens Valley (*A. b. nevadensis* at Rattlesnake Flat and *A. b. canescens* at Jawbone Canyon; Fig. 3) showed significantly different distributions (*A. b. nevadensis* [score > 0] larger than *A. b. canescens* [score < 0]; Kolmogorov-Smirnov two-sample test, $P < 0.001$), which matched results for the

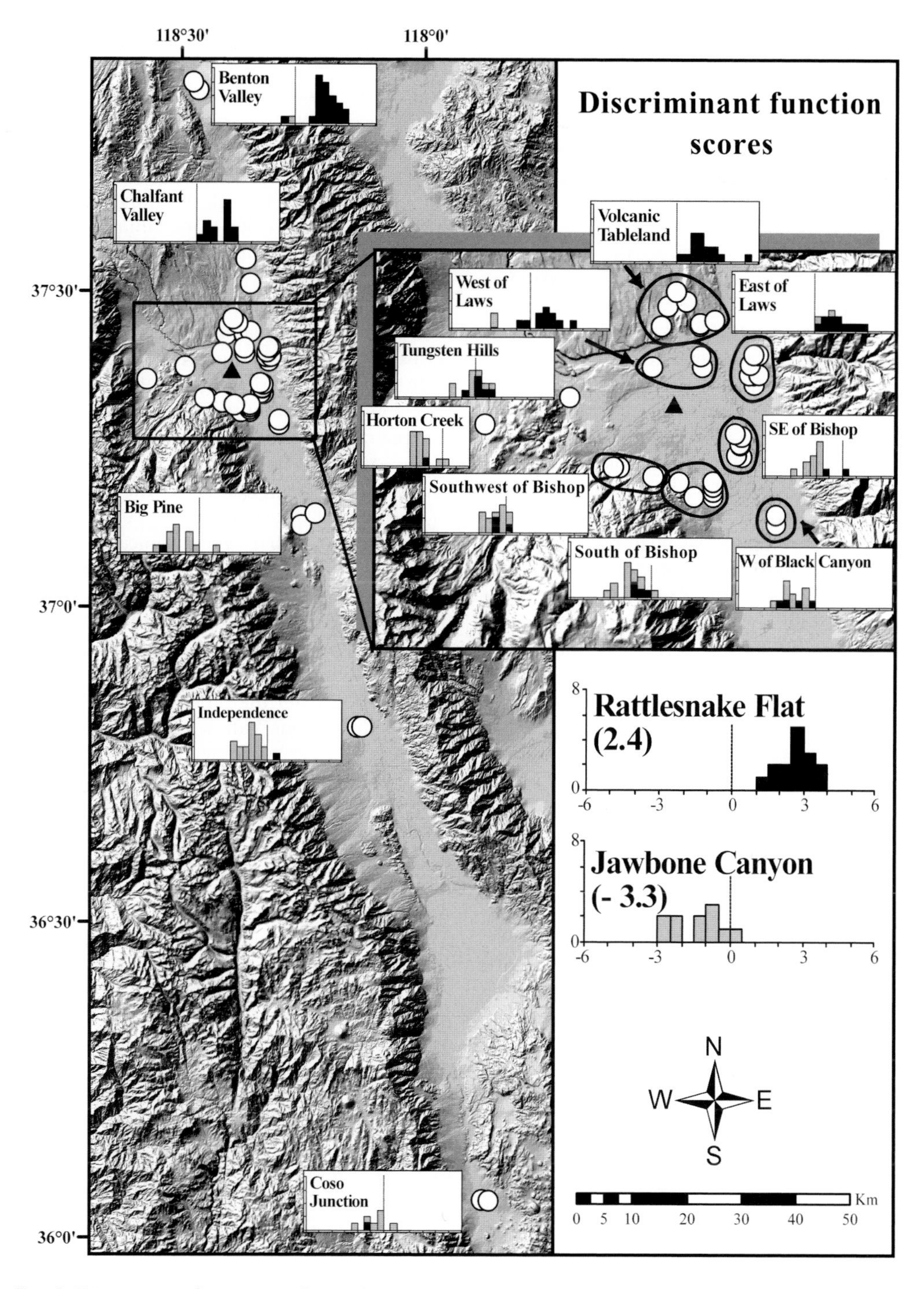


FIG. 3. Discriminant-function analysis of size characters for male *Amphispiza belli* ($n$ = 212) in the Owens Valley region, grouped by sample location (Table 1) and categorized by mtDNA haplotype group (black bars = *A. b nevadensis* [BC]; gray bars = *A. b. canescens* [AD and AC]). Discriminant-function scores of reference males from Rattlesnake Flat and Jawbone Canyon, outside the study area (see Fig. 1), were significantly different (Kolmogorov-Smirnov two-sample test, $P < 0.001$; mean scores in parentheses). Vertical bars in discriminant-function plots show a score of zero.

two subspecies across their geographic range (Cicero and Johnson 2006). Within the Owens Valley region, discriminant-function scores showed a clear trend from larger birds in the north (score > 0; e.g., Benton Valley, Chalfant Valley, Volcanic Tableland, East and West of Laws) to smaller birds in the south (score < 0; e.g., Horton Creek and Bishop south to Coso Junction). As with mtDNA, the most morphologically intermediate sample occurred at Tungsten Hills.

Analysis of mtDNA by morphology (Fig. 3) showed that most of the birds with a discriminant-function score > 0 have a haplotype in the BC group (*nevadensis*), whereas those with scores < 0 have either an AD or AC haplotype (*canescens*). Although samples from near Bishop (southwest of Bishop, south of Bishop, and west of Black Canyon) were more mixed, with relatively small birds (size typical of *A. b. canescens*) having BC haplotypes characteristic of *A. b. nevadensis*, 85.4% of individuals from the 14 core sample areas were correctly classified according to mtDNA by size. Thus, of the 212 males in the analysis, 80% of birds with a BC haplotype ($n$ = 99) were classified as BC according to size, and 90% of birds with an AD or AC haplotype ($n$ = 113) were classified by size in that group (Fig. 4). Although one would expect some recombinants to be like pure parentals, even in a hybrid swarm, it is noteworthy that most of the discriminant-function misclassifications occurred in the narrow zone where the two haplotypes meet near Bishop.

Despite the high proportion of correctly classified individuals in the region, application of a diagnosability index $D_{ij}$ for subspecies (Patten and Unitt 2002) resulted in nondiagnosable differences in wing length between *A. b. nevadensis* and *A. b. canescens* in Owens Valley ($D_{nc}$ = –1.00, $D_{cn}$ = –1.23). In this analysis, individuals were assigned to one of the two subspecies on the basis of their discriminant-function score (>0 or <0, respectively) and haplotype group (BC vs. AD/AC). Thus, potentially intermediate individuals (i.e., those with a size typical of one subspecies and haplotype characteristic of the other) were excluded. Although this result contrasts with the discriminant-function data and with the strong diagnosability between *A. b. nevadensis* and *A. b. canescens* on a broader geographic scale ($D_{nc}$ = 0.40, $D_{cn}$ = 1.01; Cicero and Johnson 2006), $D_{ij}$ relies on a single morphological trait. Furthermore, reduced diagnosability between subspecies is expected in areas of contact because of intergradation, selection, or both, in a common environment.

*Population genetic measures suggest rapid expansion into Owens Valley.*—Sequencing of haplotypes for a subset of individuals from Owens Valley revealed four unique types in *A. b. nevadensis* and nine in *A. b. canescens* (Table 2), with an average nucleotide difference of 1.4% (six haplotypes in *canescens* and 1.6% difference if individuals with the rare AC group are excluded). According to the AMOVA, nearly all the molecular variance (82.5–95.3%) is apportioned between rather than within subspecies in that region—in contrast to the pattern outside of Owens Valley, where the high variance within subspecies is attributable to the division of *A. b. canescens* into two haplotype groups (AD and AC) separated geographically (Mojave Desert vs. San Joaquin Valley, respectively; C. Cicero and N. K. Johnson unpubl. data). The higher haplotype diversity found in *A. b. canescens* is consistent with its observed nucleotide diversity, which is approximately an order of magnitude greater than that for *A. b. nevadensis* in both the Owens Valley and broader regions (Table 2).

Both Fu's (1997) $F_S$ statistics and mismatch distributions suggest rapid demographic expansion into Owens Valley, though the patterns are also consistent with a recent selective sweep in *A. b. nevadensis*. Fu's $F_S$ (Table 2) is negative for *A. b. nevadensis* (–2.369) and *A. b. canescens* (AD haplotype group, –1.908) in this region, indicating an excess of new mutations in relation to equilibrium expectations on the basis of the number of observed alleles. Similar results were obtained more broadly for *A. b. nevadensis* in the Great Basin ($F_S$ = –2.369). By contrast, an opposite positive pattern ($F_S$ = 2.963) was observed in populations of *A. b. canescens* from the Mojave Desert–San Joaquin Valley. The unimodal mismatch distributions (Fig. 5) for the two subspecies in the Owens Valley region did not differ significantly from that expected under a sudden-expansion model.

*Owens Valley transect occurs along ecological and bioclimatic gradients.*—Principal-component analysis of the 19 bioclimatic variables showed a strong latitudinal gradient among sites in the Owens Valley region (Fig. 6A)—from Coso Junction in the south (hot and dry Mojave

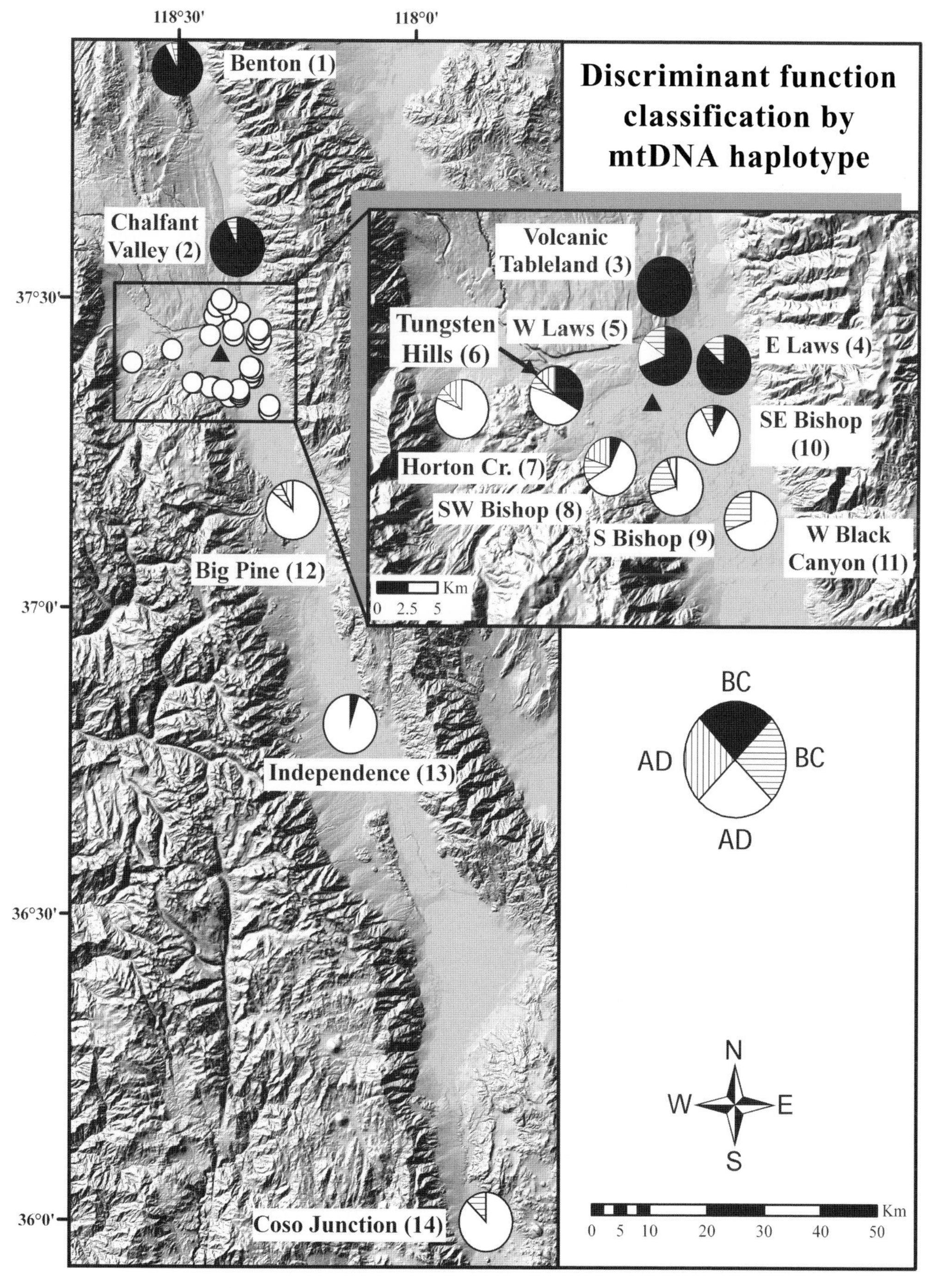


FIG. 4. Discriminant-function classification of male *Amphispiza belli* ($n$ = 212) in the Owens Valley region, grouped by sample location (Table 1) and categorized by mtDNA haplotype. Pie diagrams show the percentage of specimens with *A. b. nevadensis* (BC) or *A. b. canescens* (AD or AC) haplotypes classified correctly (solid black or white, respectively) versus incorrectly (hatched patterns)—that is, *nevadensis* haplotype classified by size as *nevadensis*, *canescens* haplotype classified by size as *canescens*. Right hatching = *canescens* haplotype classified by size as *nevadensis*. Left hatching = *nevadensis* haplotype classified by size as *canescens*.

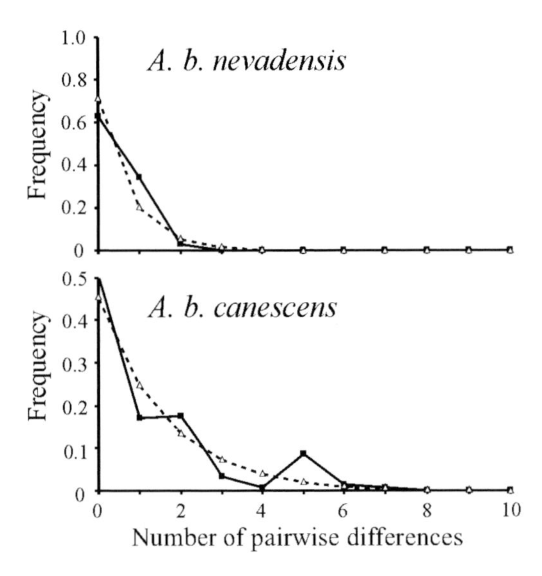


FIG. 5. Mismatch distributions for *A. b. nevadensis* and *A. b. canescens* in the Owens Valley region, on the basis of complete sequences of cytochrome *b* (closed square symbols–solid line = observed; open diamond symbols–dashed line = expected). Analysis of *A. b. canescens* excluded three individuals with sequences of the AC haplotype group, which is diagnostic for populations in the San Joaquin Valley and is rare in Owens Valley and the Mojave Desert. Observed and expected distributions are not significantly different ($P > 0.05$) and are consistent with rapid range expansions into the region.

Desert) to Benton Valley in the north (cold and wet Great Basin Desert). Localities in the contact zone clustered together toward the colder and wetter end of the gradient. In the analysis, the first two principal-component axes contained 88.4% of the variation and reflect overall temperature and precipitation conditions (PC1) versus winter and seasonal variation (PC2), respectively (Table 3). Benton Valley and Chalfant Valley (*A. b. nevadensis*) were most similar along PC1 (Fig. 6B), whereas Coso Junction and Independence (*A. b. canescens*) clustered near each other on PC2. The two sites closest to the contact zone (Chalfant Valley and Big Pine) were ecologically similar to those within the zone near Bishop. Interestingly, Horton Creek at the west end of the contact zone is an outlier with regard to climate, with PC1 being most similar to Benton Valley and PC2 reflecting conditions similar to those in Coso Junction. Individuals at that site are primarily *A. b. canescens* according to haplotype and morphology (Figs. 3 and 4), but occupy habitat of sagebrush and bitterbrush (*Purshia tridentata*) similar to that of *A. b. nevadensis*.

Ecological-niche models of *A. b. nevadensis* and *A. b. canescens* across their geographic ranges (C. Cicero and N. K. Johnson unpubl. data) provide more insight into the bioclimatic transition across Owens Valley (Fig. 7). Sites with high suitability for *A. b. nevadensis* range from Benton Valley south through Bishop and the contact zone, with Big Pine occurring at the southern end of suitable bioclimate; Independence and Coso Junction are unsuitable. On the other hand, sites with highest suitability for *A. b. canescens* include Coso Junction and Independence. According to these models, Horton Creek is unsuitable for either form. Importantly, there is essentially no overlap in Owens Valley between predicted models on the basis of bioclimatic variables.

Whereas the bioclimatic data point to a transition south of Bishop, with contact-zone sites situated in areas more suitable for *A. b. nevadensis*, the major landcover classes (WHR10) show a slightly different picture (Fig. 7). According to those layers, the vegetation makes a transition from "desert" (e.g., alkali desert scrub) to "shrub" (e.g., sagebrush) between Benton and Chalfant valleys, which is consistent with our personal field experience. Thus, sites from Chalfant Valley south to Coso Junction occur mostly in scrubby habitat more characteristic of *A. b. canescens* than *A. b. nevadensis*. However, those on the west slopes of Owens Valley (just above the valley floor) occur in an area of interdigitization between "desert" and "shrub" classes. The most notable exception again is Horton Creek, which falls purely in "shrub" comparable to habitat in Benton Valley.

## DISCUSSION

Studies of contact zones provide critical insight into evolutionary processes and barriers to sympatry between differentiated taxa (Cicero 2004, Swenson 2006). Because such zones often occur in environmental ecotones, an important goal of these studies should be to understand population-level patterns of variation in the context of climatic and ecological gradients. By combining molecular, morphological, and environmental data to examine areas of contact, distributional limits and evolutionary barriers to gene exchange between taxa become readily

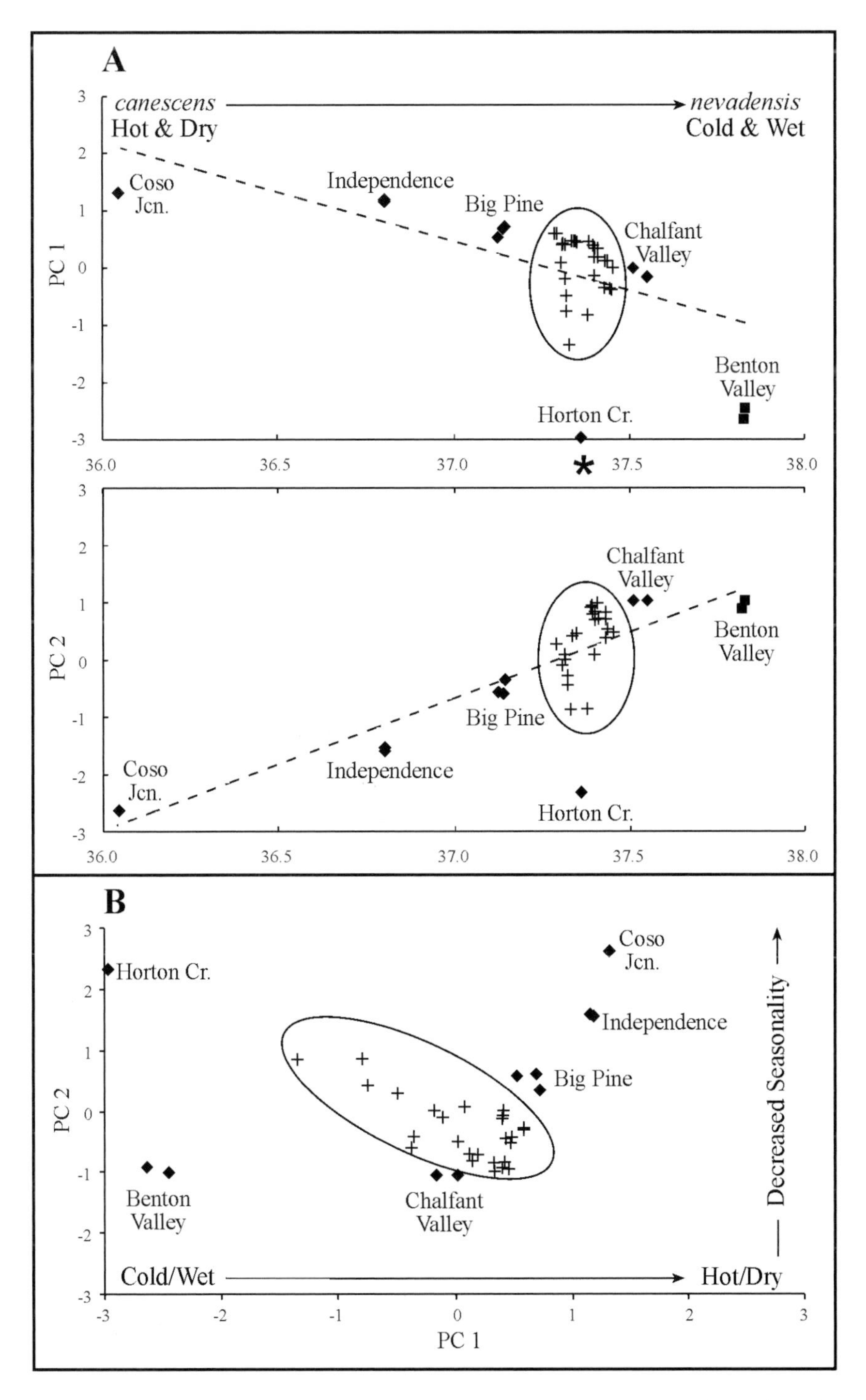


FIG. 6. Principal component scores for sample localities using 19 bioclimatic variables (Table 3) for analysis. (A) Sample localities plotted against latitude ($x$-axis) for PC1 and PC2. (B) Sample localities plotted for PC1 against PC2. In all graphs, sites along the transect are labeled with sample names (Table 1); those within the ellipsoid and shown by plus (+) symbols are in the area of contact between *A. b. canescens* and *A. b. nevadensis* near Bishop, Inyo County, California (latitude of Bishop indicated by asterisk).

TABLE 3. Factor loadings for 19 bioclimatic variables on the first three principal-component (PC) axes, based on analysis of sites along the Owens Valley transect. See text for definitions of variables.

| | Bioclimatic variable | PC1 | PC2 | PC3 |
|---|---|---|---|---|
| Bio1 | Annual mean temperature | 0.957 | 0.277 | 0.056 |
| Bio2 | Mean diurnal range | 0.203 | –0.874 | –0.437 |
| Bio3 | Isothermality | –0.142 | –0.723 | –0.634 |
| Bio4 | Temperature seasonality | 0.887 | –0.148 | 0.356 |
| Bio5 | Maximum temperature of warmest month | 0.986 | –0.126 | –0.081 |
| Bio6 | Minimum temperature of coldest month | 0.716 | 0.674 | 0.128 |
| Bio7 | Temperature annual range | 0.498 | –0.824 | –0.226 |
| Bio8 | Mean temperature of wettest quarter | 0.845 | 0.473 | 0.089 |
| Bio9 | Mean temperature of driest quarter | 0.900 | 0.098 | –0.367 |
| Bio10 | Mean temperature of warmest quarter | 0.969 | 0.220 | 0.102 |
| Bio11 | Mean temperature of coldest quarter | 0.919 | 0.378 | –0.048 |
| Bio12 | Annual precipitation | –0.783 | 0.579 | –0.214 |
| Bio13 | Precipitation of wettest month | –0.615 | 0.722 | –0.310 |
| Bio14 | Precipitation of driest month | –0.936 | –0.235 | –0.094 |
| Bio15 | Precipitation seasonality | 0.297 | 0.871 | –0.292 |
| Bio16 | Precipitation of wettest quarter | –0.636 | 0.724 | –0.262 |
| Bio17 | Precipitation of driest quarter | –0.904 | –0.215 | 0.357 |
| Bio18 | Precipitation of warmest quarter | –0.833 | –0.107 | 0.533 |
| Bio19 | Precipitation of coldest quarter | –0.640 | 0.717 | –0.269 |
| Eigenvalue | 11.086 | 5.724 | 1.741 | |
| Total variance explained (%) | | 58.3 | 30.1 | 9.2 |

apparent (e.g., Dessauer et al. 2000, Cicero 2004). Furthermore, these data can be used to infer whether contact zones associated with environmental gradients are best explained as the most likely sites of sympatry between differently adapted parental taxa (Arnold 1997, Case and Taper 2000) or as a balance between dispersal and selection across the gradient (Barton and Hewitt 1985).

Owens Valley is an ideal setting for the study of avian contact zones for several reasons: (1) it is at the western end of the Great Basin, which is well known as an area of approach or contact in numerous avian taxa (Johnson 1978); (2) the valley is situated steeply between two sets of mountains—the Sierra Nevada to the west and the White-Inyo ranges to the east—and thus any areas of contact must be confined to a north–south geographic gradient; and (3) two major ecological regions (Great Basin and Mojave Desert) meet at the northern end of the valley. Results of this study show that *A. b. nevadensis* and *A. b. canescens* occur in narrow secondary contact where these ecological zones undergo transition. Furthermore, this contact appears to be fairly recent, as indicated by indirect evidence for rapid population expansion into Owens Valley and by paleobotanical evidence for lowland woodland habitat in both deserts during recent glacial cycles (Koehler and Anderson 1994, Cicero 1996 and references therein, Woolfenden 1996). Especially interesting is the finding that populations of *A. b. canescens* outside the valley show a signature of a stable or declining rather than expanding population (positive value for Fu's $F_S$), whereas *A. b. nevadensis* shows evidence of expansion throughout its range. Thus, the Mojave Desert may be acting as a source of individuals dispersing northward into Owens Valley toward its transition with the Great Basin.

The morphological and molecular data, as well as the sharpness (~10 km) and location of the contact zone, suggest that mtDNA introgression between *A. b. nevadensis* and *A. b. canescens* is concentrated toward the northern end of Owens Valley, near the breeding limits of the two subspecies. Furthermore, *nevadensis* mtDNA haplotypes appear to be more common in *canescens*-sized birds than vice versa (i.e., *canescens* haplotypes in *nevadensis*-sized birds) in this area. This pattern may be explained by directional interbreeding of *nevadensis* females with *canescens* males, selection for smaller size in *nevadensis* as an adaptation to hotter and drier conditions, or both. Difference in timing of breeding is also an important consideration. Reproductive data show that *A. b. canescens*

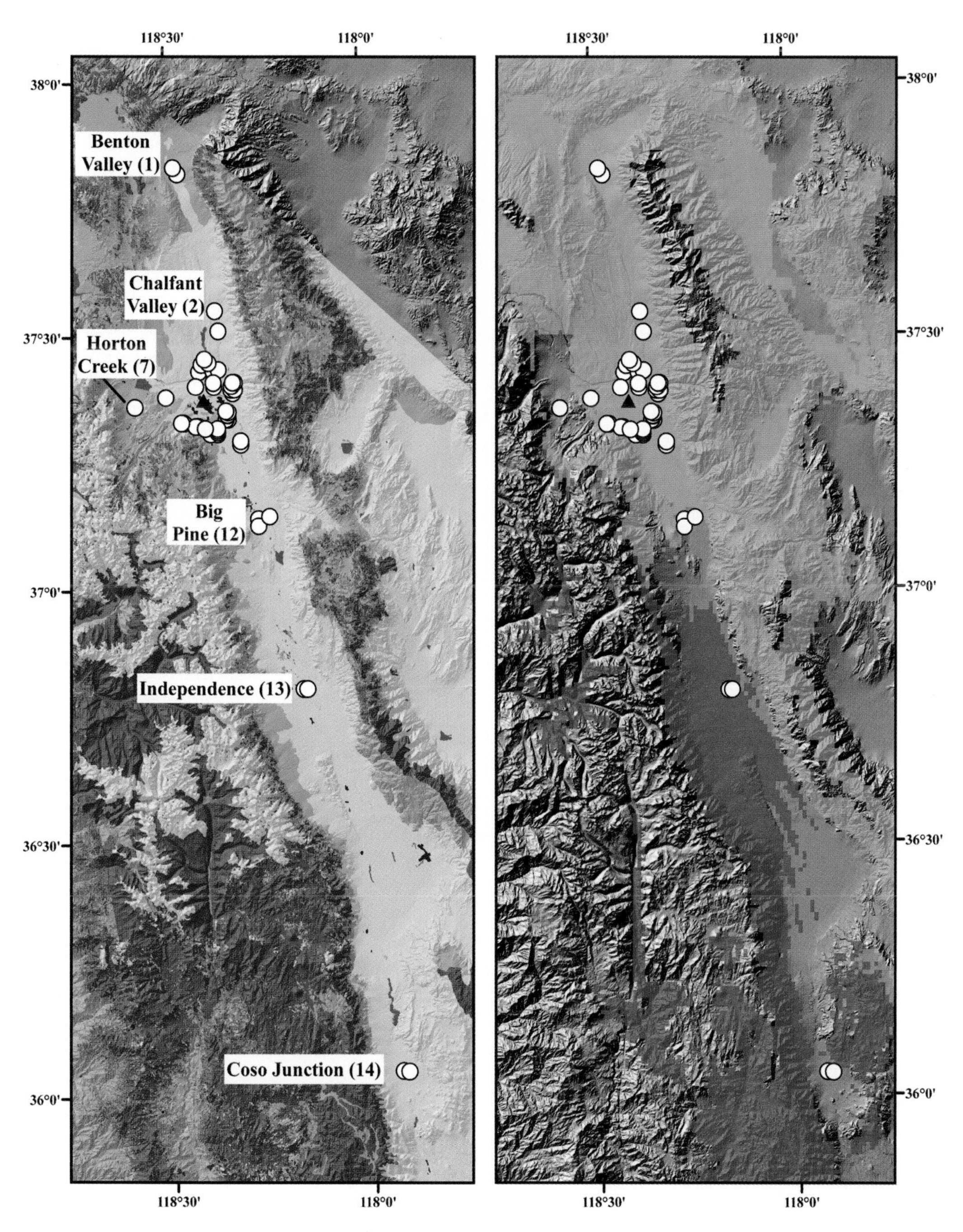


FIG. 7. (Left) Distribution of major landcover classes of wildlife habitat relationship (WHR10) for the Owens Valley transect (see text). Lightest green = "desert" (e.g., alkali desert shrub), darker green = "shrub" (e.g., sagebrush), and darkest green = conifer (not suitable for Sage Sparrows); tan = barren or agriculture. This layer is available only for California. (Right) Ecological niche models predicted for *A. b. nevadensis* (blue) and *A. b. canescens* (red) within the Owens Valley transect. Predictions are based on an analysis of bioclimatic variables and specimen points across the geographic ranges of each taxon, respectively (see text; C. Cicero and N. K. Johnson unpubl. data). Open circles in both panels show sample locations for the study (Table 1 and Fig. 2). Black triangle indicates the location of Bishop, California.

breeds earlier than *A. b. nevadensis* and that the latter subspecies often migrates northward through the range of actively breeding *canescens* (Cicero and Johnson 2006). Although this provides opportunities for interbreeding between *nevadensis* and *canescens,* such occurrences seem rare until *nevadensis* approaches the northern valley, where a steep gradient in climate and habitat favors *A. b. nevadensis* to the north and *A. b. canescens* to the south. Phenological differences may also explain the pattern at Horton Creek, where seasonal conditions typical of the breeding range of *canescens* (PC2 bioclimatic score similar to Coso Junction) likely favor earlier arrival by that subspecies, which would then start defending territories before the spring arrival of *nevadensis*.

The narrow contact zone between *A. b. nevadensis* and *A. b. canescens* at the northern end of Owens Valley is consistent with separate histories of isolation, adaptation, and genetic drift in distinctive vegetation–climate regions (e.g., see Cicero 2004). However, ecological and climatic transitions are not completely concordant in this region, with changes in landcover (sagebrush to desert scrub) apparent in Benton Valley, Mono County, whereas predicted distributional limits of *nevadensis* and *canescens* using bioclimate occur just south of Bishop, Inyo County. This difference in habitat near Benton Valley may be the reason that Grinnell and Miller (1944) surmised intergradation in that region. The observation that *nevadensis* occupies habitat more typical of *canescens* in the vicinity of Bishop, and that *canescens* occurs in habitat characteristic of *nevadensis* at Horton Creek, suggests that climate rather than vegetation type *per se* may be more important in influencing breeding occurrences. On the other hand, northernmost populations of *canescens* near Bishop experience bioclimates more similar to those of *nevadensis* in the Great Basin than to those of other populations in the Mojave Desert. This tension between suitable climate and habitat, and the ability of both forms to adapt locally to such marginal conditions, likely limits the geographic extent of contact and overlap (e.g., see Cicero 2004, Swenson 2006).

Whether *A. b. nevadensis* and *A. b. canescens* breed sympatrically can be inferred by examining the taxonomic assignment of individual birds on the basis of both mtDNA and morphology. Away from the contact zone, evidence for limited sympatry is found at Benton Valley (one breeding *A. b. canescens,* MVZ 180128) and Independence (one breeding *A. b. nevadensis,* 173336). In addition, exploratory sampling east of Owens Valley at Palmetto Wash, Esmeralda County, Nevada, revealed one *nevadensis* (MVZ 173787) in a breeding population of mostly *canescens* (MVZ 165425–165427, 173767, 173788–173790, 177065–177066; $n$ = 9). Sympatry within the contact zone is most evident west of Laws, where two unequivocal *canescens* (MVZ 177045, 178084) were collected in a sample otherwise dominated by *nevadensis* (see Fig. 3). If free interbreeding were occurring, one would expect more admixture of individuals (misclassification by morphology and mtDNA haplotype) across the region of eastern California and western Nevada where *nevadensis* and *canescens* potentially come to contact. Instead, the data clearly show diagnosable forms that are connected by sharp clines in morphology and mtDNA, and these clines are strongly correlated with climate and habitat where the Great Basin and Mojave Desert meet narrowly in the northern Owens Valley. The analysis of vocal differences between these same individuals (C. Cicero and N. K. Johnson unpubl. data) will provide additional data on their diagnosability in this region.

Although *A. b. nevadensis* and *A. b. canescens* do not appear to be freely interbreeding, the close tracking of morphology and mtDNA with ecology suggests that they may be experiencing extrinsic selection that could maintain narrow clines even in the face of extensive hybridization. The misclassification of a proportion of birds (~15%) across the zone suggests either some hybridization or size differences maintained by selection, or both. Because patterns of morphological variation can be equivocal in regard to hybridization across contact zones (Sattler and Braun 2000), determining the extent of interbreeding will require additional study of either nuclear genetic data or mated pairs, or both. Collection of mated pairs is difficult, because females are often sitting on nests and difficult to observe during the core breeding season, whereas males are singing and more visible. Nuclear genetic data can provide indirect evidence for underlying levels of gene flow, but such analysis requires a large number of loci, and especially nondiagnostic ones that are not under selection (Sattler and Braun 2000, Brumfield et al. 2001). Further study of *A. belli* will incorporate vocal and nuclear genetic data to unequivocally address the question of whether *A. b. nevadensis* and *A. b. canescens* should be regarded as biological species, *contra* previous treatments that have wrongly lumped the two forms on the basis of phenotype (Rising 1996, Patten and Unitt 2002; but see Cicero and Johnson 2006). Likewise, ongoing study (C. Cicero and N. K. Johnson unpubl. data) is addressing the issue of relationships between these subspecies and coastal populations (especially *A. b. belli*).

## Acknowledgments

The California Department of Fish and Game, U.S. Fish and Wildlife Service, and University of California Berkeley Animal Care and Use Committee issued the necessary permits and protocols for collection of specimens. C. Marchis performed the DNA laboratory work. R. Hijmans, M. Koo, W. Monahan, J. Fang, and K. Yamamoto assisted with the georeferencing and coordinate validation, GIS analyses, and preparation of maps. M. Braun, M. Matocq, and J. V. Remsen, Jr., provided useful comments on an earlier draft of this manuscript. We are grateful to all of these individuals for their assistance. All specimens, tissues, and recordings in this study are deposited in the collections of the Museum of Vertebrate Zoology, University of California, Berkeley (mvz.berkeley.edu). The WORLDCLIM data set is available at www.worldclim.org/bioclim.htm. Landcover data from the California Department of Forestry Fire and Resource Assessment Program is available at frap.cdf.ca.gov.

## Literature Cited

Arnold, M. L. 1997. Natural Hybridization and Evolution. Oxford University Press, New York.

Barton, N. H., and G. M. Hewitt. 1985. Analysis of hybrid zones. Annual Review of Ecology and Systematics 16:113–148.

Brumfield, R. T., R. W. Jernigan, D. B. McDonald, and M. J. Braun. 2001. Evolutionary implications of divergent clines in an avian (*Manacus*: Aves) hybrid zone. Evolution 55:2070–2087.

Case, T. J., and M. L. Taper. 2000. Interspecific competition, environmental gradients, gene flow, and the coevolution of species borders. American Naturalist 155:583–605.

Cicero, C. 1996. Sibling species of titmice in the *Parus inornatus* complex (Aves: Paridae). University of California Publications in Zoology, no. 128.

Cicero, C. 2004. Barriers to sympatry between avian sibling species (Paridae: *Baeolophus*) in secondary contact. Evolution 58:1573–1587.

Cicero, C., and N. K. Johnson. 2001. Higher-level phylogeny of New World vireos (Aves: Vireonidae) based on sequences of multiple mitochondrial DNA genes. Molecular Phylogenetics and Evolution 20:27–40.

Cicero, C., and N. K. Johnson. 2006. Diagnosability of subspecies: Lessons from Sage Sparrows (*Amphispiza belli*) for analysis of geographic variation in birds. Auk 123:266–274.

Dessauer, H. C., C. J. Cole, and C. R. Townsend. 2000. Hybridization among western whiptail lizards (*Cnemidophorus trigris*) in southwestern New Mexico: Population genetics, morphology, and ecology in three contact zones. Bulletin of the American Museum of Natural History 246:1–148.

Elith, J., C. H. Graham, R. P. Anderson, M. Dudik, S. Ferrier, A. Guisan, R. J. Hijmans, F. Huettmann, J. R. Leathwick, A. Lehmann, and others. 2006. Novel methods improve prediction of species' distributions from occurrence data. Ecography 29:129–151.

Excoffier, L., G. Laval, and S. Schneider. 2005. ARLEQUIN, version 3.0: An integrated software package for population genetics data analysis. Evolutionary Bioinformatics Online 1:47–50.

Fu, Y.-X. 1997. Statistical tests of neutrality of mutations against population growth, hitchhiking, and background selection. Genetics 147:915–925.

Grinnell, J., and A. H. Miller. 1944. The distribution of the birds of California. Pacific Coast Avifauna, no. 27.

Hijmans, R. J., L., Guarino, A. Jarvis, R. O'Brien, P. Mathur, C. Bussink, M. Cruz, I. Barrantes, and E. Rojas. 2005. DIVA-GIS, version 5.2. Manual. [Online.] Available at www.diva-gis.org.

Johnson, N. K. 1978. Patterns of avian geography and speciation in the Intermountain Region. Great Basin Naturalist Memoirs 2:137–159.

Johnson, N. K. 1980. Character variation and evolution of sibling species in the *Empidonax difficilis–flavescens* complex (Aves: Tyrannidae). University of California Publications in Zoology, no. 112.

Johnson, N. K. 1994. Pioneering and natural expansion of breeding distributions in western North American birds. Pages 27–44 *in* A Century of Avifaunal Change in Western North America (J. R. Jehl, Jr., and N. K. Johnson, Eds.). Studies in Avian Biology, no. 15.

Johnson, N. K., and C. Cicero. 1991. Mitochondrial DNA sequence variability in two species of sparrows of the genus *Amphispiza*. Pages 600–610 *in* Acta XX Congressus Internationalis Ornithologici (B. D. Bell, Ed.). Congressional Trust Board, Wellington, New Zealand.

Johnson, N. K., and J. A. Marten. 1992. Macrogeographic patterns of morphometric and genetic variation in the Sage Sparrow complex. Condor 94:1–19.

Koehler, P. A., and R. S. Anderson. 1994. Full-glacial shoreline vegetation during the maximum highstand at Owens Lake, California. Great Basin Naturalist 54:142–149.

Martin, J. W., and B. A. Carlson. 1998. Sage Sparrow (*Amphispiza belli*). *In* The Birds of North America, no. 326 (A. Poole and F. Gill, Eds.). Birds of North America, Philadelphia.

Mills, L. S., and F. W. Allendorf. 1996. The one-migrant-per-generation rule in conservation and management. Conservation Biology 10: 1509–1518.

Nei, M. 1987. Molecular Evolutionary Genetics. Columbia University Press, New York.

Patten, M. A., and P. Unitt. 2002. Diagnosability versus mean differences of Sage Sparrow subspecies. Auk 119:26–35.

Phillips, S. J., R. P. Anderson, and R. E. Schapire. 2006. Maximum entropy modeling of species geographic distributions. Ecological Modeling 190:231–259.

Phillips, S. J., M. Dudik, and R. E. Shapiro. 2004. A maximum entropy approach to species distribution modeling. Pages 655–662 *in* Proceedings: 21st International Conference on Machine Learning (R. Greiner and D. Shuurmans, Eds.). AAAI Press, Menlo Park, California.

Rising, J. R. 1996. A Guide to the Identification and Natural History of the Sparrows of the United States and Canada. Academic Press, London, United Kingdom.

Rogers, A. R., and H. Harpending. 1992. Population growth makes waves in the distribution of pairwise genetic differences. Molecular Biology and Evolution 9:552–569.

Rohwer, S., E. Bermingham, and C. Wood. 2001. Plumage and mitochondrial DNA haplotype variation across a moving hybrid zone. Evolution 55:405–422.

Rozas, J., J. C. SÁnchez-DelBarrio, X. Messeguer, and R. Rozas. 2003. DnaSP, DNA polymorphism analyses by the coalescent and other methods. Bioinformatics 19:2496–2497.

Ruegg, K. C., and T. B. Smith. 2002. Not as the crow flies: A historical explanation for circuitous migration in Swainson's Thrush (*Catharus ustulatus*). Proceedings of the Royal Society of London, Series B 269:1375–1381.

Sattler, G. D., and M. J. Braun. 2000. Morphometric variation as an indicator of genetic interactions between Black-capped and Carolina chickadees at a contact zone in the Appalachian Mountains. Auk 117:427–444.

Slatkin, M. 1993. Isolation by distance in equilibrium and non-equilibrium populations. Evolution 47:264–279.

Slatkin, M., and R. R. Hudson. 1991. Pairwise comparisons of mitochondrial DNA sequences in stable and exponentially growing populations. Genetics 129:555–562.

Swenson, N. G. 2006. GIS-based niche models reveal unifying climatic mechanisms that maintain the location of avian hybrid zones in a North American suture zone. Journal of Evolutionary Biology 19:717–725.

Woolfenden, W. B. 1996. Quaternary vegetation history. Pages 47–70 *in* Sierra Nevada Ecosystem Project: Final Report to Congress, vol. II. Assessments and Scientific Basis for Management Options (D. C. Erman, Ed.). Centers for Water and Wildland Resources, University of California, Davis.

Wright, S. 1951. The genetical structure of populations. Annals of Eugenics 15:323–354.

APPENDIX. Specimens and GenBank numbers of *Amphispiza belli* examined using PCR–RFLP and sequencing of cytochrome *b*. A subset of specimens (adult males only) were analyzed morphometrically. Specimen details are available from the collections database of the Museum of Vertebrate Zoology (see Acknowledgments).

| Sample name and number [a] | | MVZ specimen number | GenBank accession numbers [b] |
|---|---|---|---|
| Ref | Rattlesnake Flat | 166948–166952, 168571–168580 | EF488686, EF488687 |
| 1 | Benton Valley | 168557–168570, 180117–180133 | EF488688, EF488689, EF488690 , EF488691 |
| 2 | Chalfant Valley | 166960–166968, 169396–169398, 178279, 181710–181719 | EF488692, EF488693 |
| 3 | Volcanic Tableland | 173310–173316, 176477, 177040–177044, 177046–177050 | EF488694, EF488695 |
| 4 | East of Laws | 177155–177159, 177466–177470, 178080–178081, 178086–178087 | EF488696, EF488697, EF488698 |
| 5 | West of Laws | 173317, 176478–176482, 177045, 177160, 177465, 178082–178085 | EF488699, EF488700, EF488701, EF488702 |
| 6 | Tungsten Hills | 178065–178079 | EF488703, EF488704, EF488705, EF488706 |
| 7 | Horton Creek | 178049–178064 | EF488707, EF488708, EF488709 |
| 8 | Southwest of Bishop | 173321–173323, 173548–173552, 177057–177064 | EF488710, EF488711, EF488712, EF488713 |
| 9 | South of Bishop | 173553–173558, 176496–176501, 177051–177056 | EF488714, EF488715, EF488716, EF488717, EF488718, EF488719 |
| 10 | Southeast of Bishop | 173318–173320, 173761–173765, 176483–176487 | EF488720, EF488721 |
| 11 | West of Black Canyon | 173324–173328, 176488–176495 | EF488722, EF488723, EF488724, EF488725 |
| 12 | Big Pine | 173329–173333, 173766–173775 | EF488726, EF488727 |
| 13 | Independence | 173334–173343, 173776–173785 | EF488728, EF488729 |
| 14 | Coso Junction | 170321–170338 | EF488730, EF488731, EF488732, EF488733, EF488734, EF488735 |
| Ref | Jawbone Canyon | 169351–169354, 170283–170293 | EF488736, EF488737, EF488738 |

[a] Sample names and numbers refer to Table 1 and Figure 1. Rattlesnake Flat and Jawbone Canyon are reference samples outside the Owens Valley region.

[b] GenBank numbers are available only for sequenced individuals ($n$ = 53). MVZ = Museum of Vertebrate Zoology, University of California, Berkeley.

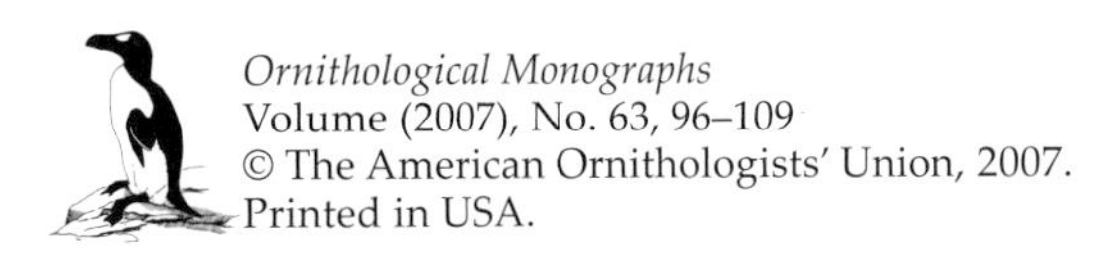
*Ornithological Monographs*
Volume (2007), No. 63, 96–109

Printed in USA.

CHAPTER 8

# STATISTICAL ASSESSMENT OF CONGRUENCE AMONG PHYLOGEOGRAPHIC HISTORIES OF THREE AVIAN SPECIES IN THE CALIFORNIA FLORISTIC PROVINCE

Kevin J. Burns,[1] Matthew P. Alexander, Dino N. Barhoum, and Erik A. Sgariglia

*Department of Biology, San Diego State University, San Diego, California 92182, USA*

Abstract.—Comparing phylogeographic histories of codistributed species can reveal how common historical events and processes have influenced lineage diversification across species. Within the California Floristic Province, the phylogeographies of a diversity of plants and animals have been studied; however, only a few bird species have been examined in this region. We compared phylogeographic histories of three species of birds—Wrentit (*Chamaea fasciata*), California Thrasher (*Toxostoma redivivum*), and White-headed Woodpecker (*Picoides albolarvatus*)—to each other as well as to the phylogeographic histories of other California taxa. Qualitatively comparing phylogeographies of these three species reveals some similarities and some differences. All three species exhibit similar levels of sequence divergence, experienced recent range expansions, and show a division between southern and northern populations in the vicinity of the Transverse Ranges. However, nested-clade phylogeographic analysis suggests that different processes have influenced current patterns of genetic structure in these three species. Other taxa within California show a similar division between northern and southern populations; however, our species did not show concordant geographic breaks elsewhere that have been identified for other species. Using a statistical assessment of concordance, our data showed more agreement across a broader regional scale than among closely spaced populations. Phylogeographic trees of the three species showing relationships among counties were not statistically congruent with each other or with a tree representing other California taxa. However, at the broader level of geomorphic province, the Wrentit and White-headed Woodpecker showed complete concordance. *Received 30 June 2006, accepted 5 February 2007.*

Resumen.—La comparación de la historia filogeográfica de las especies con distribuciones similares, puede revelar cómo los eventos históricos y los procesos comunes que han experimentado estas especies han influenciado la diversificación de sus linajes. Dentro de la provincia florística de California, se ha estudiado la filogeografia de diversas plantas y animales; sin embargo, sólo se han estudiado unas pocas especies de aves. En este estudio, comparamos la historia filogeográfica de tres especies de aves—*Chamaea fasciata, Toxostoma redivivum* y *Picoides albolarvatus*; además, comparamos la historia filogeográfica de otros taxones californianos. Las comparaciones cualitativas de la filogeografía entre las tres especies antes mencionadas, revelan algunas diferencias y similitudes. Las tres especies muestran niveles similares de divergencia genética, presentando, además, expansiones recientes en sus rangos de distribución y mostrando una división entre las poblaciones del norte y del sur en las cercanías de los rangos transversales. Sin embargo, los análisis filogeográficos de clados encajados (NCA) sugieren que los patrones de estructura genética en estas tres especies han sido determinados por procesos diferentes. Otros taxones dentro de California muestran una división similar entre las poblaciones del norte y del sur; sin embargo, nuestras especies no muestran una interrupción en otras áreas geográficas, las que han sido identificadas en especies con una misma distribución. Utilizando un análisis estadístico de correspondencia nuestras aves muestran mayor concordancia a través de una amplia escala regional que entre

[1]E-mail: kburns@sunstroke.sdsu.edu

poblaciones que esten cercanamente espaciadas. Los árboles filogeográficos de las tres especies que muestran las relaciones entre condados no son estadísticamente congruentes entre ellos, como tampoco lo son con los árboles de otros taxones de California. Sin embargo, al nivel más amplio de la provincia geomórfica, existe una total congruencia entre *Chamaea fasciata* y *Picoides albolarvatus*.

Studies of geographic variation in morphology (Johnson 1966, 1980; Johnson and Marten 1992) have long provided insight into the evolutionary history of species. With the advent of DNA sequencing, study of geographic variation in genetic data through phylogeographic analyses has provided additional insights into the processes responsible for diversification within species (Barrowclough et al. 2004, Cicero and Johnson 2007). Comparing the phylogeographic histories of multiple species provides a broader context for these diversification processes by identifying how geology, geography, ecology, and climatic history may have shaped population structure across codistributed species (Bermingham and Avise 1986, Zink 2002). However, comparing phylogeographic histories is difficult, partly because methods assessing statistical congruence across species have been explored only recently (Edwards and Beerli 2000, Calsbeek et al. 2003, Lapointe and Rissler 2005).

For strong inferences to be made, the phylogeographic histories of a large number of diverse taxa within the same region are necessary. Such a large data set now exists for the region known as the California Floristic Province. This region, which stretches from southwestern Oregon to northern Baja California, mostly coincides with the political boundaries of the state of California. The complex geological history of the area has led to a diversity of habitats and high levels of endemism, and the area is considered one of 34 biodiversity hotspots in the world (Mittermeier et al. 2005). In fact, more endemic taxa occur there than in any other similar-sized region in North America (Calsbeek et al. 2003). Among birds, Miller (1951) included 13 species and nearly 100 subspecies of birds within the Californian avifauna (defined by having their distributions confined to or at least centered within California).

Two recent studies have synthesized phylogeographic data for this region. Calsbeek et al. (2003) found largely congruent patterns of genetic diversity across 55 species (24 plants, 10 insects, 8 reptiles, 6 mammals, 5 birds, and 2 amphibians). Patterns of differentiation in many of these species corresponded to common past physical events. For example, most animal taxa studied showed an obvious genetic split dividing northern and southern populations around the Transverse Ranges. More recently, Lapointe and Rissler (2005) used statistical tests of congruence to compare phylogeographic histories of nine taxa. The species they studied included three amphibians, two mammals, one reptile, one bird, one insect, and one plant, each sampled from an average of 17 counties. They found significant phylogeographic signal in common across these taxa and, thus, were able to combine the data from each of these nine taxa into a single supertree representing biogeographic relationships among regions. In general, these studies show that different species have congruent phylogeographic histories within the California Floristic Province. However, only a handful of bird studies have been conducted in the area, compared with the large number of studies available for other taxa (Calsbeek et al. 2003). Here, we compare the phylogeographic histories of three different bird species and assess their congruence with each other and with other species in the region.

The three species we studied occur across the California Floristic Province and are generally sedentary, with little postbreeding dispersal (Baker et al. 1995, Garrett et al. 1996, Cody 1998). Two of the species, the Wrentit (*Chamaea fasciata*) and the California Thrasher (*Toxostoma redivivum*), are mostly restricted to California. The third species, the White-headed Woodpecker (*Picoides albolarvatus*), has a broader distribution that also includes parts of Oregon, Washington, Idaho, and British Columbia. The Wrentit and the California Thrasher occupy similar mid- and low-elevation scrub and chaparral habitats (Cody 1998, Geupel and Ballard 2002), whereas the White-headed Woodpecker is found at higher elevations, largely confined to coniferous areas >850 m in elevation (Garrett et al. 1996). Similarities in habitat between the Wrentit and the California Thrasher may lead to more similar phylogeographic histories, if similar climatic histories shape patterns of population structure. Alternatively, broader geographic or geological factors that transcend habitat may play a more important role, leading to congruence across all three species.

Each species was the focus of a separate phylogeographic study (Sgariglia and Burns 2003, Alexander and Burns 2006, Burns and Barhoum 2006). Thus, we do not present details of the phylogeographic history of each species. Instead, we concentrate on two main goals: (1) assessing congruence among these three species and (2) assessing congruence between each of these species and the general patterns seen in other taxa of the California Floristic Province.

Comparing phylogeographic histories among diverse organisms is often challenging because sampling designs, genetic markers, and analytical methods differ among studies, and because a rigorous methodology does not exist for identifying whether similarities and differences among species are statistically significant. We have attempted to overcome these problems by using the same genetic markers, by sampling each species from the same or nearby localities, by using the same analytical methods, and by using a statistical approach to assess congruence. We first qualitatively compared phylogeographies of each species using results of standard analyses, including analysis of molecular variation (AMOVA), mismatch distributions, haplotype networks, and nested-clade phylogeographic analysis (NCPA). Next, we used Lapointe and Rissler's (2005) method to statistically assess congruence among the different phylogeographic histories across two spatial scales.

## Methods

For each species, we sampled multiple populations throughout the distribution (Table 1) and sequenced between 1,777 and 2,148 base pairs of mitochondrial DNA (mtDNA) per individual. Data from multiple gene regions (cytochrome *b*, ATP synthase 6, and ATP synthase 8) were combined in all analyses. An AMOVA calculation of $\Phi_{st}$ using pairwise distances (Excoffier et al. 1992) provided a measure of overall genetic structure of populations for each species. The $\Phi$ statistic indicates the proportion of nucleotide variation divided among populations and ranges from 0 to 1.0, with completely subdivided populations having a $\Phi$ statistic of 1.0. The distribution of pairwise differences among individuals (the mismatch distribution) was used to infer whether a population has undergone a sudden population expansion (Rogers 1995, Rogers and Harpending 1992). Agreement between the observed and expected distribution under a sudden-expansion model was tested following Schneider and Excoffier (1999). Evidence of a population expansion was also tested using Fu's $F_s$ (Fu 1997) and Tajima's $D$ (Tajima 1989). Assuming neutrality, a significantly negative value of these statistics indicates an excess of new mutations in relation to equilibrium expectations and leads to rejection of population stasis. Relationships among individuals were illustrated using parsimony-based haplotype networks constructed using TCS, version 1.13 (Clement et al. 2000). We conducted an NCPA (Templeton 1998, 2004) on the haplotype networks to infer population-level processes. The root of the intraspecific network was inferred through neutral coalescent theory (Castelloe and Templeton 1994, Crandall et al. 1994, Clement et al. 2000). In general, the haplotype with the greatest frequency and the most connections is the most likely to be identified as the ancestral haplotype. Further details on these methods of analysis as well as detailed information on sampling localities, location of voucher specimens, and molecular laboratory methods are provided elsewhere (Sgariglia and Burns 2003, Alexander and Burns 2006, Burns and Barhoum 2006).

In addition to comparing results of these methods for each species, we also statistically tested for phylogeographic congruence following the approach outlined in Lapointe and Rissler (2005). In the first step of this approach, a tree is constructed for each species that shows relationships among predefined geographic areas (e.g., counties). Trees are constructed by recoding the original data so that individuals from the same area are pooled into common units. To compare

Table 1. Sampling coverage and intraspecific variation in the three species compared in the present study.

| Species | Number of individuals | Number of unique haplotypes | Number of populations | $P$ distance [a] | $\Phi$ statistic [b] |
|---|---|---|---|---|---|
| Wrentit | 61 | 39 | 20 | 0.41% (0–0.51%) | 0.41 |
| California Thrasher | 64 | 37 | 21 | 0.34% (0–0.77%) | 0.46 |
| White-headed Woodpecker | 78 | 25 | 24 | 0.25% (0–0.68%) | 0.33 |

[a] Mean and range of uncorrected distance presented.

[b] Values are for comparisons among populations within each species; all are significant ($P < 0.001$).

our results with those of Lapointe and Rissler (2005), we first used counties as our geographic unit for assessing phylogeographic histories at a relatively fine spatial scale (Fig. 1). Only samples within California were used to construct these trees. The average pairwise distance based on Kimura's (1980) two-parameter model between each county was computed, and this matrix was used to draw a neighbor-joining tree (Saitou and Nei 1987) for each species using PAUP*, version 4.0b10 (Swofford 2002). Each tree then was compared to each other, and the size of the maximum agreement subtree, or MAST score, was calculated in PAUP* and used as a congruence index. The MAST score is the largest possible tree compatible with a given pair of trees, and higher MAST scores indicate more congruence between trees. For example, if two trees showed identical relationships among 20 terminals, the MAST score would equal 20. Because the same counties were not always sampled in each species, MAST scores were normalized by dividing the score by the number of counties sampled in common between the two species. Thus, pairs of trees with complete congruence have a normalized MAST score of 1.0, and complete lack of agreement would lead to a score of 0. For example, if two species were both sampled in the same 20 counties, and the relationships of 18 of those counties were the same for both species, the normalized MAST score would equal 0.9. To test whether the normalized MAST score between two trees was significantly high, we compared the observed MAST score to a null distribution of MAST scores expected by comparing two random trees. To construct the null distribution, we generated 1,000 random trees in PAUP*, with the number of terminals the same as the number of counties in common between the two trees. The agreement metric "d1" was calculated between all possible pairs of random trees (499,500 total contrasts), and the distribution of d1 was transformed into a null distribution of MAST scores. Two trees were considered significantly congruent if their observed MAST score fell within the extreme upper portion of the null distribution. If trees were significantly congruent, they could be combined into a supertree (Gordon 1986, Bininda-Emonds 2004) showing relationships among all counties sampled by either data set. Using this approach, we compared the county trees of each species to those of each other species. In addition, we compared each of the trees of our species to the county supertree of California taxa presented in Lapointe and Rissler (2005).

To look for congruence on a broader spatial scale, we repeated the analysis with our bird data by dividing populations into 12 regional areas within California (Fig. 2). The areas we used correspond to the major geographic and geological features in California. Our regions are similar to the geomorphic provinces of Hill (1984), with the exception that we divided the Coast Ranges into northern Coast Ranges and southern Coast Ranges (Schoenherr 1992). The Wrentit and the California Thrasher occur in six of these geomorphic provinces, and the White-headed Woodpecker occurs in eight. Species were sampled from all geomorphic provinces in which they occur. Looking at the data divided into geographic provinces may be more biologically meaningful than comparing data on the basis of the political subdivision of county. In addition, the comparison among geomorphic provinces provides a broader spatial scale. Because birds are more vagile than most other organisms, widespread gene flow may prevent phylogeographic signal among closely spaced populations (among counties within geomorphic provinces), yet a pattern may be apparent across a broader regional scale (among geomorphic provinces).

## Results

### Intraspecific Variation

Haplotype networks, analyses of intraspecific variation, and NCPA for each species are presented in more detail elsewhere (Sgariglia and Burns 2003, Alexander and Burns 2006, Burns and Barhoum 2006); thus, these results are summarized only briefly here. All three species showed similar levels of intraspecific sequence divergence (Table 1). For all three species, mean values of percentage of sequence divergence were <1%, typical of mitochondrial DNA for intraspecific comparisons of birds (Avise and Walker 1998, Ditchfield and Burns 1998). For each species, AMOVA indicated that this variation was significantly structured among populations, with $\Phi_{st}$ values ranging from 0.33 to 0.41. Mismatch distributions for each species did not differ significantly from the expected distribution of a growing population ($P$ = 0.17–0.86), indicating that each species likely experienced a recent range expansion. Negative values of Fu's $F_s$ and Tajima's $D$ were obtained for each species, providing further evidence of recent range expansions.

### Comparison of Haplotype Networks

For each species, haplotype network construction resulted in a single network in which all connections fell within a 95% plausible set of relationships. All three networks (not shown) were completely nested within four-step clades, indicating a similar temporal framework for each species. The California Thrasher network

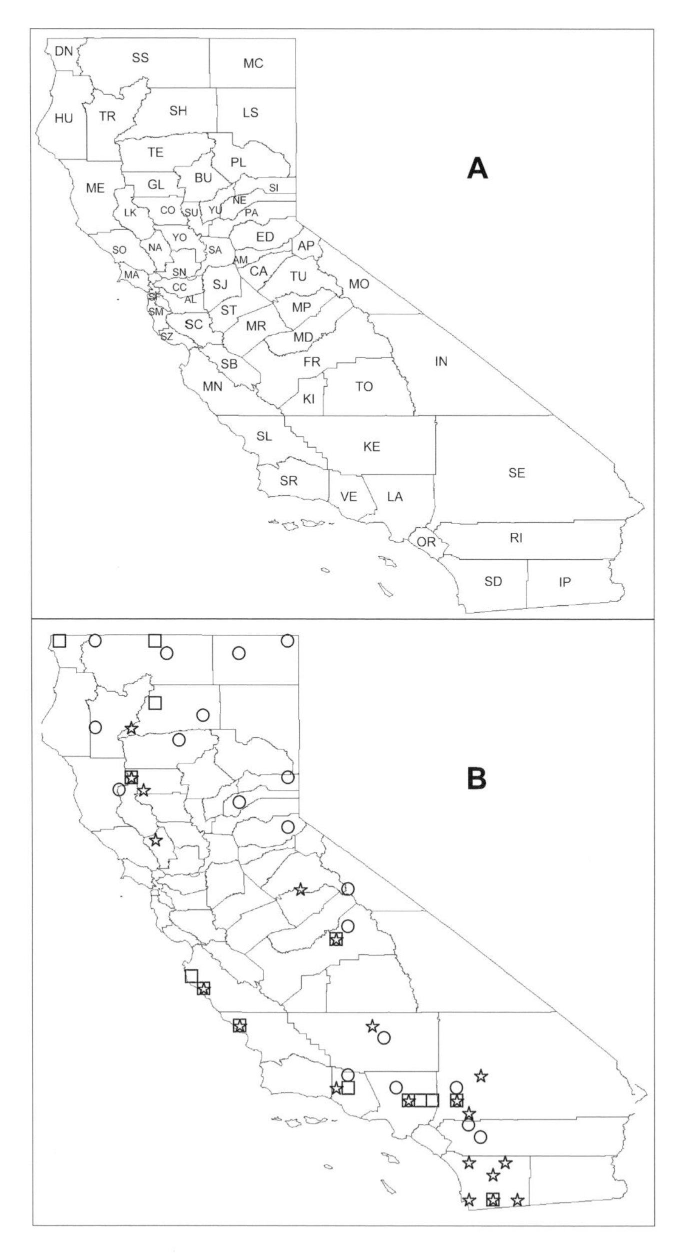


Fig. 1. (A) Map of California showing counties. Abbreviations of counties: AM = Amador, AP = Alpine, BU = Butte, CA = Calaveras, CC = Contra Costa, CO = Colusa, DN = Del Norte, ED = El Dorado, FR = Fresno, GL = Glenn, HU = Humboldt, IN = Inyo, IP = Imperial, KE = Kern, KI = Kings, LA = Los Angeles, LK = Lake, LS = Lassen, MA = Marin, MC = Modoc, MD = Madera, ME = Mendocino, MN = Monterey, MO = Mono, MP = Mariposa, MR = Merced, NA = Napa, NE = Nevada, OR = Orange, PA = Placer, PL = Plumas, RI = Riverside, SA = Sacramento, SB = San Benito, SC = Santa Clara, SD = San Diego, SE = San Bernardino, SF = San Francisco, SH = Shasta, SI = Sierra, SJ = San Joaquin, SL = San Luis Obispo, SM = San Mateo, SN = Sonoma, SO = Solano, SR = Santa Barbara, SS = Siskiyou, ST = Stanislaus, SU = Sutter, SZ = Santa Cruz, TE = Tehama, TO = Tuolumne, TR = Trinity, TU = Tulare, VE = Ventura, YO = Yola, and YU = Yuba. (B) General location of sampling sites within California for all three species. Specific locality data for each species is provided (see text). Stars indicate sites where California Thrashers were sampled, squares indicate sites where Wrentits were sampled, and circles indicate sites where White-headed Woodpeckers were sampled.

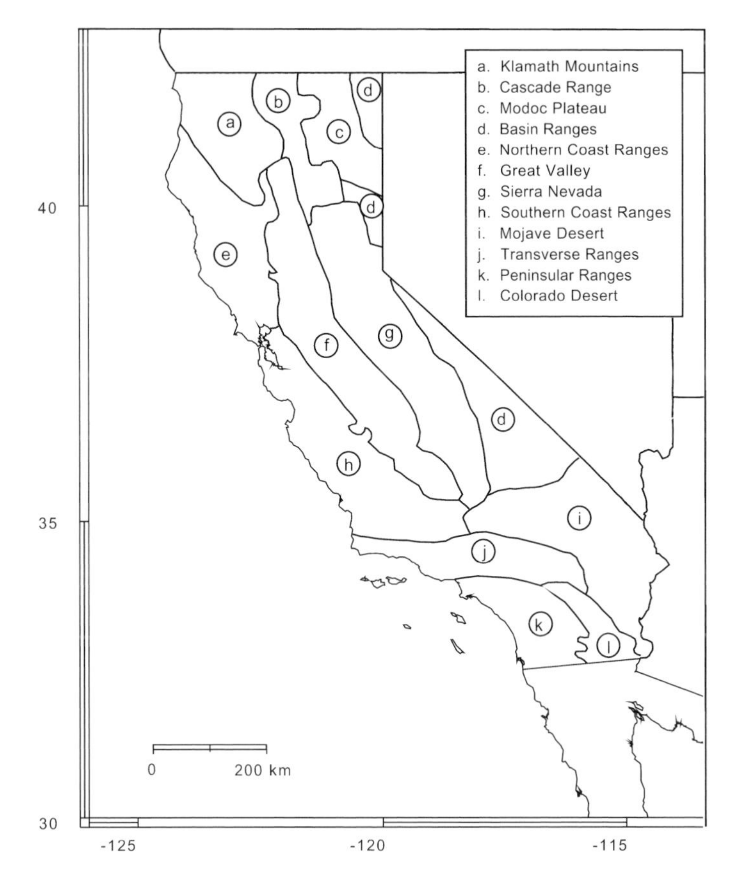


FIG. 2. Geomorphic provinces used to explore relationships among regions within the California Floristic Province.

was characterized by nine unsampled haplotypes that separated most southern haplotypes from those in the north, reflecting a comparatively greater amount of genetic divergence observed between the northern and southern populations. By contrast, sampled haplotypes in the networks of the Wrentit and White-headed Woodpecker were mostly connected to each other by other sampled haplotypes. Few geographically concordant clades are readily apparent by comparing haplotype networks of each species. However, all three species identify a separation of southern California birds from those in the rest of the network. The geographic boundaries of these southern California clades differed slightly in each of the three species. For the California Thrasher and the White-headed Woodpecker, the separation of northern and southern birds occurred just north of the Transverse Ranges, with birds found in the Peninsular and Transverse ranges in a clade separate from other regions (Fig. 2). For the Wrentit, the separation occurred within the Transverse Ranges, with individuals in the San Bernardino Mountains (a mountain range within the Transverse Ranges) clustering with those from the Peninsular Ranges. The oldest haplotype of both the Wrentit and California Thrasher networks occurred in the same geographic area. For both networks, the oldest haplotype was found in several populations spanning the southern Coast Ranges and the Transverse Ranges. By contrast, for the White-headed Woodpecker, the oldest haplotype was found in the northern part of the distribution.

### NESTED-CLADE PHYLOGEOGRAPHIC ANALYSIS

We identified, by NCPA, several clades in each species with significant values of genetic and geographic distance for which inferences about evolutionary processes could be made.

The evolutionary history of each species was complex and characterized by a diversity of processes, including range expansion, restricted gene flow with isolation by distance, and allopatric fragmentation (Table 2). These processes occurred at different times and in different regions for each species. Geographic and genetic relationships among haplotypes at the earliest branching point in the networks indicate that different processes influenced the earliest evolutionary events in each species. The inferred process at the deepest level of the haplotype network is range expansion for the Wrentit, allopatric fragmentation for the California Thrasher, and restricted gene flow for the White-headed Woodpecker. Although NCPA identified a clade of southern California birds for each species, the processes influencing this division were not the same for each species. For the California Thrasher and the Wrentit, NCPA inferred allopatric fragmentation. However, for the White-headed Woodpecker, range expansion by long-distance colonization was inferred (Table 2).

### Comparison of Regional Trees

Trees built using pairwise distances between counties (Fig. 3) show few similarities to each other. However, all three species show a separation between northern and southern California counties. In the California Thrasher tree, a long branch connects the counties of these two regions. This corresponds to the large number of unsampled haplotypes connecting northern and southern California populations seen in the haplotype network of this species. Although the county trees of the Wrentit and the White-headed Woodpecker (Fig. 3) also show a separation between northern and southern California counties, this separation is not marked by a branch length longer than others seen in the network. Other than the separation of northern and southern counties, the three networks do not show many similarities in county relationships, and their normalized MAST scores are not significantly congruent (Table 3). Thus, the three trees are, overall, no more similar to each other than would be expected at random and cannot be combined together into a supertree. In addition, none of our avian trees shows significant congruence to the multispecies county network of Lapointe and Rissler (2005). Thus, each species we studied shows a significantly different phylogeographic pattern in relation to each other as well as to the other California taxa at the spatial level of counties.

Constructing trees for each species on the basis of geomorphic provinces (Fig. 4) also showed a division between southern and northern California regions. For the Wrentit and the California Thrasher, the Transverse Ranges, southern Coast Ranges, and Peninsular Ranges are in a clade separate from the northern Coast Ranges, Klamath Mountains, and the Sierra Nevada. The White-headed Woodpecker does not occur in the southern Coast Ranges, but still shows a separation of Peninsular Ranges and Transverse Ranges from the geomorphic provinces of northern California. Although a north-versus-south split is shared among all three species, the trees show different relationships among regions within northern California and southern California. Different topologies within northern California prevent the California Thrasher tree from being significantly congruent with the White-headed Woodpecker tree (Fig. 4 and Table 3). Likewise, different topologies within both northern and southern California prevent significant congruence between the California Thrasher tree and the Wrentit tree (Fig. 4 and Table 3). However, both the Wrentit and the White-headed Woodpecker are completely congruent for the regions in which they both occur. Therefore, they were combined into a supertree (Fig. 5) showing relationships among eight different regions within the California Floristic Province.

## Discussion

### Qualitative Assessment of Phylogeographic Congruence

Using standard approaches (AMOVA, mismatch distributions, haplotype networks, and NCPA) to compare phylogeographic histories across the distribution of each species indicates both similarities and differences among species. The similar level of genetic divergence found in each species suggests similar timing of diversification within each lineage, assuming equal molecular rates of evolution across these three bird lineages and a lack of error associated with ancestral polymorphism (Edwards and Beerli 2000). All have experienced recent range

TABLE 2. Inferences obtained from NCPA for clades with statistically significant distance values. Each line indicates results for a particular clade in which inferences could be made. Asterisks identify the clade and the inferred process responsible for the genetic differences observed between northern and southern populations.

| Species | Step level [a] | Inferred process | Geographic region of inferred process |
|---|---|---|---|
| Wrentit | 4 | Range expansion (long-distance and contiguous) | Expansion from Peninsular and Southern Coast ranges to remainder of the distribution |
| | 3* | Allopatric fragmentation* | Separation of Peninsular Ranges and San Bernardino mountains (part of the Transverse Ranges) from the rest of the Transverse and Southern Coast ranges |
| | 3 | Restricted gene flow with isolation by distance | Across Klamath Mountains, Northern Coast Ranges, and Sierra Nevada |
| California Thrasher | 4* | Allopatric fragmentation* | Transverse and Peninsular ranges mostly separated from remainder of the distribution |
| | 3 | Range expansion (contiguous) | Expansion from southern Sierra Nevada to all other parts of the distribution |
| | 3 | Restricted gene flow with isolation by distance | Across Transverse and Peninsular ranges |
| | 1 | Allopatric fragmentation | Part of Southern Coast Ranges separated from Transverse and Peninsular ranges |
| White-headed Woodpecker | 4 | Restricted gene flow with isolation by distance | Across entire distribution |
| | 3* | Range expansion (long-distance colonization)* | Dispersal to Transverse and Peninsular ranges from remainder of the distribution |
| | 3 | Restricted gene flow with isolation by distance | Across entire distribution except for Peninsular Ranges |
| | 2 | Restricted gene flow with isolation by distance | Across Sierra Nevada, Basin, Cascade, and Northern Coast ranges |
| | 1 | Allopatric fragmentation | Basin Ranges separated from Northern Coast and Cascade ranges |

[a] Step level reflects relative age of clades within each cladogram. For each species, 1-step clades are the youngest and 4-step clades are the oldest.

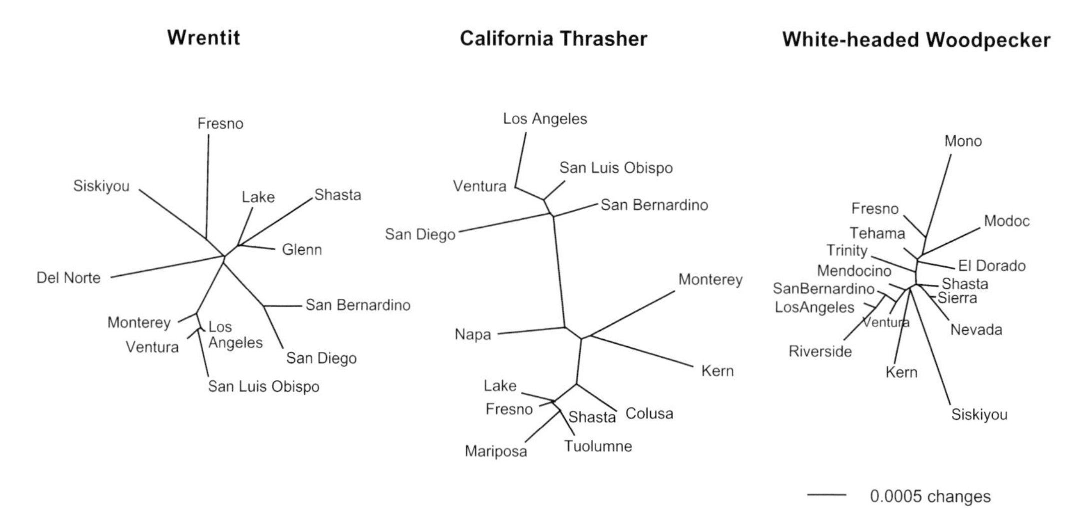


FIG. 3. Trees representing relationships among counties. Trees are unrooted, and branch lengths are proportional to amount of genetic divergence.

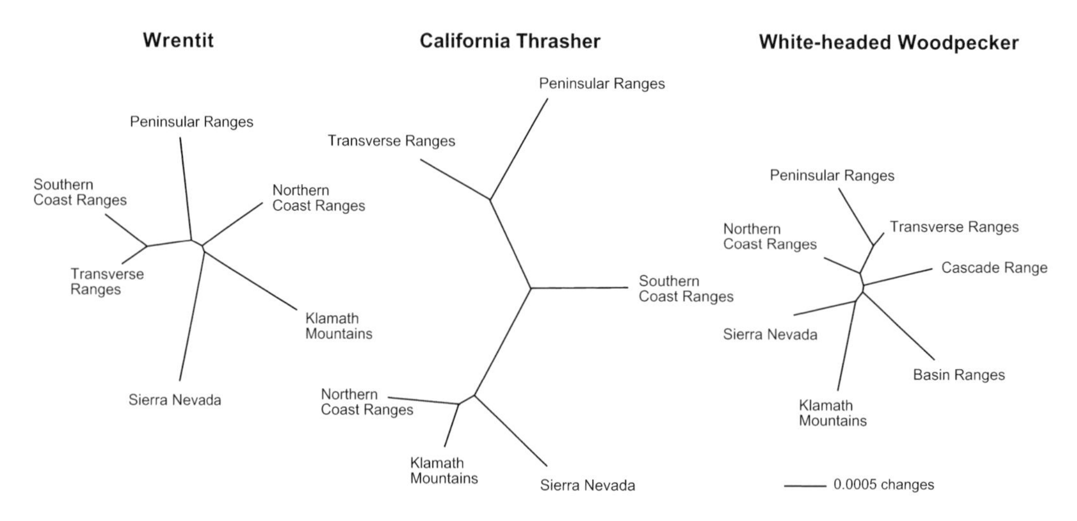


FIG. 4. Trees representing relationships among geomorphic provinces. Trees are unrooted, and branch lengths are proportional to amount of genetic divergence.

expansions, as reflected in their unimodal mismatch distributions and the NCPA inferences of range expansion for large parts of the distribution. However, other than range expansion, NCPA identified few processes that influenced the three species similarly at the same time and space. A split between northern and southern populations was identified, but the precise boundaries and processes influencing this split were different in each species. Inferred ancestral origin also showed some level of agreement across taxa. Both the Wrentit and the California Thrasher experienced a southern ancestry followed by a northward postglacial range expansion. However, the White-headed Woodpecker likely had a more northern ancestry.

A qualitative comparison of phylogeographic histories of the species studied here and those of codistributed taxa also show similarities and differences. Calsbeek et al. (2003) analyzed genetic data from 55 species of plants and animals, and they identified several geographic barriers that seemed to influence multiple species in the same way. Most animals displayed

TABLE 3. Scores representing assessment of pairwise comparison of county-based trees and geomorphic-province-based trees.

| Spatial scale of tree | Trees compared | Number of regions shared | MAST score | Normalized MAST Score | $P$ value |
|---|---|---|---|---|---|
| County | Wrentit vs. California Thrasher | 9 | 5 | 0.56 | 0.97 |
| | Wrentit vs. White-headed Woodpecker | 6 | 4 | 0.67 | >0.99 |
| | California Thrasher vs. White-headed Woodpecker | 6 | 5 | 0.83 | 0.26 |
| | Lapointe and Rissler (2005)[a] vs. Wrentit | 11 | 6 | 0.55 | 0.70 |
| | Lapointe and Rissler (2005)[a] vs. California Thrasher | 13 | 6 | 0.46 | 0.95 |
| | Lapointe and Rissler (2005)[a] vs. White-headed Woodpecker | 13 | 7 | 0.54 | 0.35 |
| Geomorphic province | Wrentit vs. California Thrasher | 6 | 4 | 0.67 | >0.99 |
| | Wrentit vs. White-headed Woodpecker | 5 | 5 | 1.00 | 0.07 |
| | California Thrasher vs. White-headed Woodpecker | 5 | 4 | 0.80 | 0.87 |

[a] Refers to county supertree for California taxa reported by Lapointe and Rissler (2005).

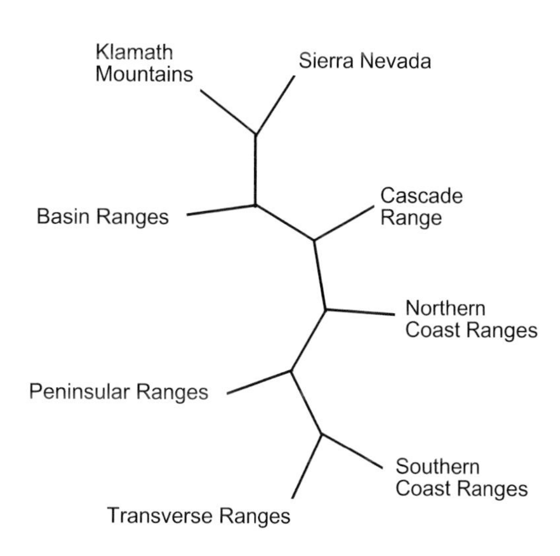


FIG. 5. Supertree showing combined relationships among geomorphic provinces for the Wrentit and the White-headed Woodpecker. Tree is unrooted, and branch lengths are constrained to be of equal length.

a genetic split dividing populations on either side of the Transverse Ranges. In addition, multiple species also were affected by population genetic breaks in the Sierra Nevada, Coast Ranges, Monterey Bay, and the Los Angeles Basin. However, Calsbeek et al. (2003) noted that the few bird species included in their study were relatively weakly differentiated and did not show breaks around any of these barriers. By contrast, our study shows that some birds have responded to the Transverse Ranges in a manner similar to that of nonavian species. All three species showed a genetic break in this region, though the process influencing this break likely differs among species (Table 2). In addition, the Oak Titmouse (*Baeolophus inornatus*), a species not included in the Calsbeek et al. (2003) study but included in the Lapointe and Rissler (2005) study, also shows a genetic break between northern and southern populations in the vicinity of the Transverse Ranges (Cicero 1996). Other than the Transverse Ranges, we did not find similar breaks in the other regions identified as important barriers by Calsbeek et al. (2003). Timing of differentiation around the Transverse Ranges appears to be generally concordant among our bird species and other species previously studied. Calsbeek et al. (2003; Table 1) provided dates of divergence for six taxa in this region. For four of these, divergence times date to within or slightly more than 1 mya. Assuming a molecular clock rate of 1.6–2.0% divergence per million years for bird mtDNA (Shields and Wilson 1987, Fleischer et al. 1998; but see García-Moreno 2004), differentiation among our species in this region also occurred within the last million years. The pattern of southern ancestry inferred for the Wrentit and California Thrasher has also been identified in a number of other California vertebrates (Smith 1979, Tan and Wake 1995, Cicero 1996, Rodríguez-Robles et al. 1999, Maldonado et al. 2001, Matocq 2002, Richmond and Reeder 2002). However, from just the above qualitative description, strong overall conclusions about generality of phylogeographic patterns between our birds and other taxa cannot be made.

### Statistical Assessment of Congruence

Using Lapointe and Rissler's (2005) method, the three species we studied showed incongruent histories when compared with each other at the relatively fine spatial scale of counties. More congruence was found at a broader regional scale. The geomorphic-province trees for the Wrentit and White-headed Woodpecker were completely congruent, despite the different habitat requirements and elevational preferences of these two species. Although the White-headed Woodpecker and the California Thrasher phylogeographies were not statistically congruent, they share relationships of four of the five regions in common. However, with only five regions in common between these two species, statistical power is low for this comparison.

Lapointe and Rissler's (2005) test of congruence among nine taxa reached a conclusion similar to that of Calsbeek et al. (2003), namely that species within the California Floristic Province are characterized by congruent phylogeographic histories. However, the three species in our study showed little agreement with the multispecies supertree of Lapointe and Rissler (2005). In general, for the counties in common between our species and those of the Lapointe and Rissler (2005) tree, only half of these counties agreed in topological arrangement between trees. This suggests that haplotypes are shared across a broader area in our bird species than in the species included in Lapointe and Rissler's (2005) supertree. Lapointe and Rissler (2005) did not include a tree based on broader regional areas; therefore, we cannot statistically compare our geomorphic-province tree to a Lapointe and Rissler (2005) regional supertree. However, visual inspection of the Lapointe and Rissler (2005) county tree shows that counties in the Transverse Ranges, Peninsular Ranges, and southern Coast Ranges are grouped together, which is similar to our geomorphic province supertree. In addition, counties in the Sierra Nevada, Klamath Mountains, Cascade Range, and northern Coast Ranges also group together in agreement with our supertree. However, the tree of Lapointe and Rissler (2005) shows counties in the Basin Ranges grouping with the Peninsular Range counties, whereas our supertree places the Basin Ranges adjacent to regions in northern California. Our results suggest that avian species may show more congruent phylogeographic histories at regional levels than at a fine scale within the California Floristic Province. Greater dispersal abilities of birds compared with many other taxa are likely responsible for this difference. More movement among closely spaced populations would result in lower genetic differences, obscuring the ability to differentiate fine-scale geographic patterns. In addition, the ability of birds to undergo distributional shifts over hundreds of kilometers in the short period of a few decades (Johnson 1994) would also obscure phylogenetic signal. Additional studies of bird species are still needed to help clarify the extent to which the phylogeographic histories of birds comply with the congruent patterns seen in other species in this region.

The approach of Lapointe and Rissler (2005) is an improvement over previous methods of assessing congruence, in that a statistical test is employed so that common geographic associations can be identified with confidence. In addition, once regional associations are identified, these can be tested for differences in climatic or other variables (Lapointe and Rissler 2005). However, identifying congruence among areas using this approach does not reveal the evolutionary processes underlying an association. For example, a tree may indicate that two areas are grouped together, but this association could be attributable to vicariant events in the past as well as to ongoing, restricted gene flow. In addition, the congruence test does not incorporate branch lengths; thus, temporal differences and relative support for an association are ignored. For example, the California Thrasher showed greater genetic divergence between northern and southern populations than the other two species, which suggests that events causing the split between these two regions may have occurred at different time-scales in the three species. Because this and other kinds of important information are not considered when assessing topological congruence, we advocate the continued use of traditional phylogeographic methods in conjunction with the supertree approach of Lapointe and Rissler (2005). The two approaches are complementary, with the former providing detailed information about the timing and evolutionary process of each species and the latter revealing common geographic associations across a region.

### Acknowledgments

Funds were provided by the American Museum of Natural History (Frank M. Chapman Memorial Fund), Los Angeles Audubon Society (Ralph W. Schreiber Ornithology Research Award), and the National Science Foundation (0217817, 0315416). We thank A. Bohonak for calculation of the null distribution and other statistical advice. The U.S. Fish and Wildlife Service and California Department of Fish and Game provided collecting permits. For additional tissues, we thank the Natural History Museum of Los Angeles County, the Museum of Vertebrate Zoology at the University of California Berkeley, the Burke Museum of Natural History and Culture, the Louisiana State University Museum of Natural Science Collection of Genetic Resources, the American Museum of Natural History, the San Diego Natural History Museum, P. Escalante, R. Frenzel, and J. Dudley. For comments on the manuscript, we thank C. Cicero, J. V. Remsen, and two anonymous reviewers.

### Literature Cited

Alexander, M. P., and K. J. Burns. 2006. Intraspecific phylogeography and adaptive divergence in the White-headed Woodpecker. Condor 108:489–518.

Avise, J. C., and D. Walker. 1998. Pleistocene phylogeographic effects on avian populations and the speciation process. Proceedings of the Royal Society of London, Series B 265:457–463.

Baker, M., N. Nur, and G. R. Geupel. 1995. Correcting biased estimates of dispersal and survival due to limited study area: Theory and application using Wrentits. Condor 97: 663–674.

Barrowclough, G. F., J. G. Groth, L. A. Mertz, and R. J. Gutiérrez. 2004. Phylogeographic structure, gene flow and species status in Blue Grouse (*Dendragapus obscurus*). Molecular Ecology 13:1911–1922.

Bermingham, E., and J. C. Avise. 1986. Molecular zoogeography of freshwater fishes in the southeastern United States. Genetics 113: 939–965.

Bininda-Emonds, O. R. P. 2004. The evolution of supertrees. Trends in Ecology and Evolution 19:315–322.

Burns, K. J., and D. N. Barhoum. 2006. Population-level history of the Wrentit (*Chamaea fasciata*): Implications for comparative phylogeography in the California Floristic Province. Molecular Phylogenetics and Evolution 38:117–129.

Calsbeek, R., J. N. Thompson, and J. E. Richardson. 2003. Patterns of molecular evolution and diversification in a biodiversity hotspot: The California Floristic Province. Molecular Ecology 12:1021–1029.

Castelloe, J., and A. R. Templeton. 1994. Root probabilities for intraspecific gene trees under neutral coalescent theory. Molecular Phylogenetics and Evolution 3:102–113.

Cicero, C. 1996. Sibling species of titmice in the *Parus inornatus* complex (Aves: Paridae). University of California Publications in Zoology, no. 128.

Cicero, C., and N. K. Johnson. 2007. Narrow contact of desert sage sparrows (*Amphispiza belli nevadensis* and *A. b. canescens*) in the Owens Valley, eastern California: Evidence from mitochondrial DNA, morphology, and GIS-based niche models. Pages 78–95 *in* Festschrift for Ned K. Johnson: Geographic Variation and Evolution In Birds (C. Cicero and J. V. Remsen, Jr., Eds.). Ornithological Monographs, no. 63.

Clement, M., D. Posada, and K. A. Crandall. 2000. TCS: A computer program to estimate gene genealogies. Molecular Ecology 9: 1657–1659.

Cody, M. L. 1998. California Thrasher (*Toxostoma redivivum*). *In* The Birds of North America, no. 323 (A. Poole and F. Gill, Eds.). Birds of North America, Philadelphia.

Crandall, K. A., A. R. Templeton, and C. F. Sing. 1994. Intraspecific phylogenetics: Problems and solutions. Pages 273–297 *in* Models of Phylogeny Reconstruction (R. W. Scotland, D. J. Siebert, and D. M Williams, Eds.). Clarendon Press, Oxford, United Kingdom.

Ditchfield, A. D., and K. J. Burns. 1998. DNA sequences reveal phylogeographic similarities of Neotropical bats and birds. Journal of Comparative Biology 3:165–170.

Edwards, S. V., and P. Beerli. 2000. Perspective: Gene divergence, population divergence, and the variance in coalescence time in phylogeographic studies. Evolution 54:1839–1854.

Excoffier, L., P. E. Smouse, and J. M. Quattro. 1992. Analysis of molecular variance inferred from metric distances among DNA haplotypes: Application to human mitochondrial DNA restriction data. Genetics 131:479–491.

Fleischer, R. C., C. E. McIntosh, and C. L. Tarr. 1998. Evolution on a volcanic conveyor belt: Using phylogeographic reconstructions and K-Ar-based ages of the Hawaiian Islands to estimate molecular evolutionary rates. Molecular Ecology 7:533–545.

Fu, Y.-X. 1997. Statistical tests of neutrality of mutations against population growth, hitchhiking and background selection. Genetics 147:915–925.

García-Moreno, J. 2004. Is there a universal mtDNA clock for birds? Journal of Avian Biology 35:465–468.

Garrett, K. L., M. G. Raphael, and R. D. Dixon. 1996. White-headed Woodpecker (*Picoides albolarvatus*). *In* The Birds of North America, no. 252 (A. Poole and F. Gill, Eds.). Academy of Natural Sciences, Philadelphia, and American Ornithologists' Union, Washington, D.C.

Gordon, A. D. 1986. Consensus supertrees: The synthesis of rooted trees containing overlapping sets of labeled leaves. Journal of Classification 3:335–348.

Geupel, G. R., and G. Ballard. 2002. Wrentit (*Chamaea fasciata*). *In* The Birds of North America, no. 654 (A. Poole and F. Gill, Eds.). Birds of North America, Philadelphia.

Hill, M. 1984. California Landscape: Origin and Evolution. University of California Press, Berkeley, California.

Johnson, N. K. 1966. Morphologic stability versus adaptive variation in the Hammond's Flycatcher. Auk 83:179–200.

Johnson, N. K. 1980. Character variation and evolution of sibling species in the *Empidonax difficilis–flavescens* complex (Aves: Tyrannidae). University of California Publications in Zoology, no. 112.

Johnson, N. K. 1994. Pioneering and natural expansion of breeding distributions in western North American birds. Pages 27–44 *in* A Century of Avifaunal Change in Western North America (J. R. Jehl, Jr., and N. K. Johnson, Eds.). Studies in Avian Biology, no. 15.

Johnson, N. K., and J. A. Marten. 1992. Macrogeographic patterns of morphometric and genetic variation in the Sage Sparrow complex. Condor 94:1–19.

Kimura, M. 1980. A simple method for estimating evolutionary rate of base substitutions through comparative studies of nucleotide sequences. Journal of Molecular Evolution 16: 111–120.

Lapointe, F.-J., and L. J. Rissler. 2005. Congruence, consensus, and the comparative phylogeography of codistributed species in California. American Naturalist 166:290–299.

Maldonado, J. E., C. Vilà, and R. K. Wayne. 2001. Tripartite genetic subdivisions in the ornate shrew (*Sorex ornatus*). Molecular Ecology 10: 127–147.

Matocq, M. D. 2002. Phylogeographical structure and regional history of the dusky-footed woodrat, *Neotoma fuscipes*. Molecular Ecology 11:229–242.

Miller, A. H. 1951. An analysis of the distribution of the birds of California. University of California Publications in Zoology 50: 531–644.

Mittermeier, R. A., P. R. Gil, M. Hoffman, J. Pilgrim, T. Brooks, C. G. Mittermeier, J. Lamoreux, and G. A. B. da Fonseca. 2005. Hotspots Revisited: Earth's Biologically Richest and Most Endangered Terrestrial Ecoregions. Conservation International, Washington, D.C.

Richmond, J. Q., and T. W. Reeder. 2002. Evidence for parallel ecological speciation in scincid lizards of the *Eumeces skiltonianus* species group (Squamata: Scincidae). Evolution 56: 1498–1513.

Rodríguez-Robles, J. A., D. F. Denardo, and R. E. Staub. 1999. Phylogeography of the California mountain kingsnake, *Lampropeltis zonata* (Colubridae). Molecular Ecology 8:1923–1934.

Rogers, A. R. 1995. Genetic evidence for a Pleistocene population explosion. Evolution 49:606–615.

Rogers, A. R., and H. Harpending. 1992. Population growth makes waves in the distribution of pairwise genetic differences. Molecular Biology and Evolution 9:552–569.

Saitou, N., and M. Nei. 1987. The neighbor-joining method: A new method for reconstruction phylogenetic trees. Molecular Biology and Evolution 4:406–425.

Schneider, S., and L. Excoffier. 1999. Estimation of demographic parameters from the distribution of pairwise differences when the mutation rates vary among sites: Application to human mitochondrial DNA. Genetics 152:1079–1089.

Schoenherr, A. A. 1992. A Natural History of California. University of California Press, Berkeley.

Sgariglia, E. A., and K. J. Burns. 2003. Phylogeography of the California Thrasher (*Toxostoma redivivum*) based on nested clade analysis of mitochondrial DNA variation. Auk 120:346–361.

SHIELDS, G. F., AND A. C. WILSON. 1987. Calibration of mitochondrial DNA evolution in geese. Journal of Molecular Evolution 24:212–217.

SMITH, M. F. 1979. Geographic variation in genetic and morphological characters in *Peromyscus californicus*. Journal of Mammalogy 60:705–722.

SWOFFORD, D. L. 2002. PAUP*: Phylogenetic Analysis Using Parsimony (*and Other Methods), version 4.0b10. Sinauer Associates, Sunderland, Massachusetts.

TAJIMA, F. 1989. Statistical method for testing the neutral mutation hypothesis by DNA polymorphism. Genetics 123:585–596.

TAN, A. M., AND D. B. WAKE. 1995. Mitochondrial DNA phylogeny of the California newt *Taricha torosa* (Caudata, Salamandridae). Molecular Phylogenetics and Evolution 4:383–394.

TEMPLETON, A. R. 1998. Nested clade analyses of phylogeographic data: Testing hypotheses about gene flow and population history. Molecular Ecology 7:381–397.

TEMPLETON, A. R. 2004. Statistical phylogeography: Methods of evaluating and minimizing inference errors. Molecular Ecology 13:789–809.

ZINK, R. M. 2002. Methods in comparative phylogeography, and their application to studying evolution in the North American aridlands. Integrative and Comparative Biology 42: 953–959.

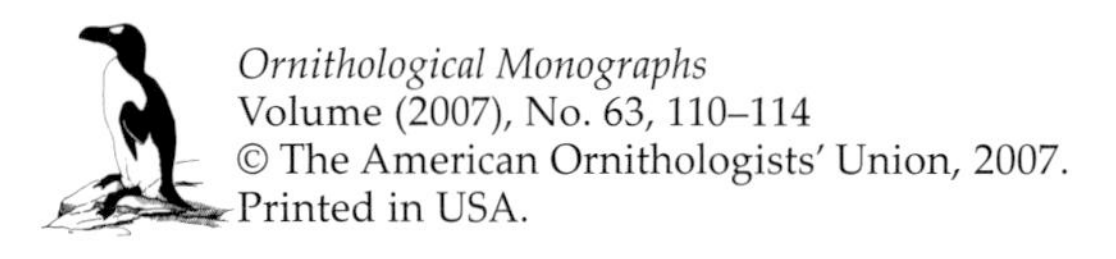
*Ornithological Monographs*
Volume (2007), No. 63, 110–114


CHAPTER 9

# THE "WALKING EAGLE" *WETMOREGYPS DAGGETTI* MILLER: A SCALED-UP VERSION OF THE SAVANNA HAWK (*BUTEOGALLUS MERIDIONALIS*)

Storrs L. Olson[1]

*Division of Birds, National Museum of Natural History, P.O. Box 37012, Smithsonian Institution, Washington, D.C. 20013, USA*

Abstract.—The so-called "walking eagle," currently known as *Wetmoregyps daggetti* from the Pleistocene of southern California and northern Mexico, is practically identical in morphology and proportion to the living Savanna Hawk (*Buteogallus meridionalis*) but ~40% larger. It should therefore be known as *Buteogallus daggetti*, new combination. Its habits were perhaps like those of a Savanna Hawk in that it was capable of taking much larger prey, given that the weight of *B. daggetti* may have exceeded that of the modern Secretarybird (*Sagittarius serpentarius*), and it may have occupied a similar niche. Any connection between its extinction and the disappearance of the North American mammalian megafauna is dubious at best. *Received 27 November 2006, accepted 5 February 2007.*

Resumen.—*Wetmoregyps daggetti*, encontrado en el Pleistoceno en el sur de California y norte de México, es prácticamente idéntico en morfología y en proporciones al gavilán pita venado *Buteogallus meridionalis*, aunque un 40% más grande. Por lo tanto, debe considerarse como *Buteogallus daggetti*. Probablemente, sus hábitos eran muy similares a los del gavilán pita venado, siendo capaz de capturar grandes presas. Su tamaño excede al del secretario (*Sagittarius serpentarius*), pudiendo ocupar un nicho similar. Cualquier conexión entre la extinción de *Wetmoregyps daggetti* y la desaparición de la megafauna de mamíferos en Norte America es, cuando menos, dudosa

In the course of investigating the relationships of several species of large fossil raptors from Cuba with William Suárez, I borrowed skeletons of the two living species of "eagles" of the genus *Harpyhaliaeetus*, which are extremely rare in collections. Taking the opportunity to study their relationships in turn, we collaborated with Stephen Parry and also included in our comparisons some of the large species of Accipitridae described from the Pleistocene of Rancho La Brea, California. We eventually realized that we could identify a previously unrecognized radiation of raptors, all of which could be accommodated in the genus *Buteogallus*. Three of the living species of *Buteogallus*, all differing in skeletal proportions, appeared to have nearly identical counterparts, either living or fossil, that were scaled up more than 30–40% in size, so that they have been regarded as "eagles." Each of these instances will be treated separately, beginning here with the fossil species now known as *Wetmoregyps daggetti*.

That species was described originally by Miller (1915) as *Morphnus daggetti*, on the basis of a very long, slender tarsometatarsus that Miller regarded as having belonged to a "walking eagle." Three additional tarsometatarsi and an incomplete tibiotarsus were assigned to the species a few years later (Miller 1925). Miller (1928, 1931) identified fragmentary tarsometatarsi from the Carpinteria asphalt deposits as this species. In re-evaluating its affinities, he had "no hesitation in placing the Pleistocene bird nearer to *Urubitinga* [= *Buteogallus*] than to *Morphnus*," while noting the "strong superficial resemblance of the Daggett Eagle to…weak-footed and small-mouthed raptors," such as *B. urubitinga* or *Caracara* spp. (Miller 1928:255). Nevertheless, probably influenced by its great size, he created for it a new genus, *Wetmoregyps*, though the allusion to vultures in the name

[1]E-mail: olsons@si.edu

(Greek, *gyps* = vulture), as we shall see, was misleading. It continued to be known as *W. daggetti* up to the present.

Apart from referring a coracoid, and tentatively some phalanges, Howard (1932:15) could "add almost nothing to the knowledge of this species," and she gave no measurements. She later mentioned two additional individuals of *W. daggetti* from Rancho La Brea but did not indicate which elements had come to light (Howard 1936). The geographical range of the species was extended >2,100 km to the southeast with the discovery of two coracoids and a tarsometatarsus in San Josecito Cave, Nuevo León, Mexico (Miller 1943).

Miller's (1928:255) diagnosis of *W. daggetti* was as follows:

> Size equal to, or greater than, *Aquila*; tarsus extremely elongate; papilla of tibialis anticus placed high up on the shaft, resulting in a ratio of power arm over weight arm of approximately 12.5 per cent; superficial resemblance to *Urubitinga urubitinga*, but less excavated on the antero-proximal face and with trochleae more nearly of equal size and elevation.
>
> Additional points of divergence from *U. urubitinga* include the following: (1) Inner cotyla exceeds the outer to a greater degree; (2) sagittal diameter of head greater in relation to transverse diameter; (3) outer hypotarsal ridge lower but broader; (4) scar of the distal rudiment of metatarsal 1 shorter and placed centrad from the inner profile of the bone; distal foramen placed lower down.

Howard (1932) found that fragmentary remains of *W. daggetti* were difficult to discern from like elements of the fossil eagle then known as "*Morphnus*" *woodwardi*, which is understandable because, as it turns out, *M. woodwardi* is another member of the buteogalline radiation referred to above (S. Olson unpubl. data).

Miller and Howard thus came very close to discerning the true relationships of the "Daggett Eagle," and had they extended their comparisons, they doubtless would have noted the extreme similarity between *W. daggetti* and the living Savanna Hawk (*Buteogallus meridionalis*; formerly placed in the monotypic genus *Heterospizias*). Howard (1932) had borrowed a skeleton of *B. meridionalis* and included its measurements in her monograph, but she evidently did not notice its similarity to *W. daggetti*.

## Systematics

Family Accipitridae

Genus *Buteogallus* Lesson, 1830

Synonym. *Wetmoregyps* Miller, 1928. 255; type-species by original designation *Morphnus daggetti* Miller, **new synonymy**.

*Buteogallus daggetti* Miller (1915), **new combination**

*Morphnus daggetti* Miller, 1915:179; 1925:97.

*Wetmoregyps daggeti*: Miller, 1928:255.

*Referred material examined.*—Complete left tarsometatarsus LACM K3159; left tibiotarsus lacking proximal articulation LACM 79744.

*Comparative skeletal material examined (all USNM).*—*Buteogallus meridionalis* 32968, 319439, 319440, 347849, 560138, 622379, 630248; *B. urubitinga* 343972, 345786; *Sagittarius serpentarius* 621021, 621022.

*Comparisons.*—Once the comparison is made between *B. daggetti* and *B. meridionalis* (Figs. 1–3 and Table 1), there is little left to be said. Apart from size, the tarsometatarsi are practically identical. The size of *B. daggetti* is ~40% greater than the average size of *B. meridionalis*. Considering that the mass of the larger bird would have been more than triple that of the smaller, the lack of more purely size-related differences between them is remarkable. The individual of *B. meridionalis* with the largest skeleton examined in the series weighed 1,050 g, whereas *B. daggetti* likely exceeded 3,000 g (see below). The greatest proportional difference is in the width of the shaft of the tibiotarsus, which is 49% larger in *B. daggetti*, versus 38–41% in the other width measurements.

The only noticeable qualitative difference is the protuberance in *B. daggetti* on the lateral surface of the proximal end, in the area identified by Baumel et al. (1979, their fig. 14A,C) as the *sulcus musculo fibularis longus* and *impressio ligamentum collateralis lateralis*. This protuberance is not observed in *B. meridionalis*. The tarsometatarsus of *B. daggetti* agrees with that of *B. meridionalis* and differs from that of *B. urubitinga* in being longer and more gracile, with a shorter medial hypotarsal crest (shortness of this crest is characteristic of all the buteogallines) and in having a larger and more distally situated distal foramen (as noted by Miller 1928). There are certainly no differences that could be considered of

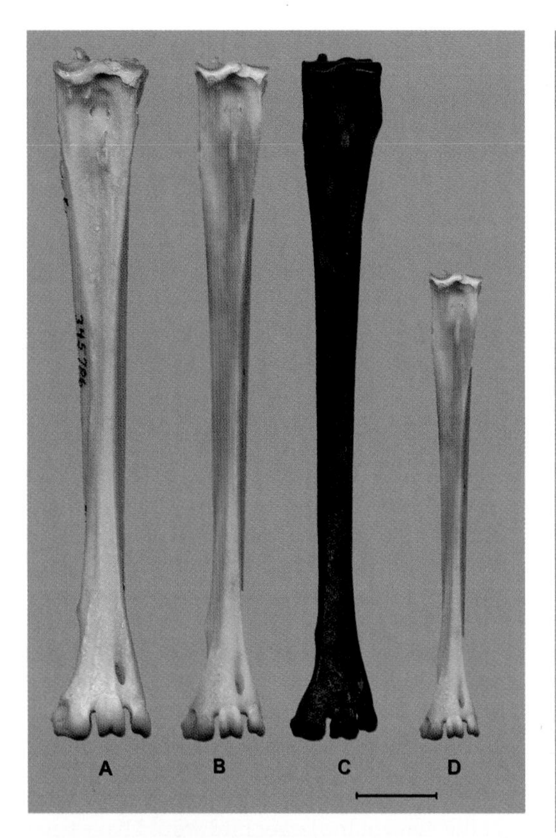


FIG. 1. Left tarsometatarsi of *Buteogallus* in anterior view: (A) *B. urubitinga* USNM 345786, (B) *B. meridionalis* USNM 630248, (C) *B. daggetti* LACM K3159, and (D) *B. meridionalis* USNM 630248. (A) and (B) are enlarged to the same size as (C). Scale = 2 cm for (C) and (D).

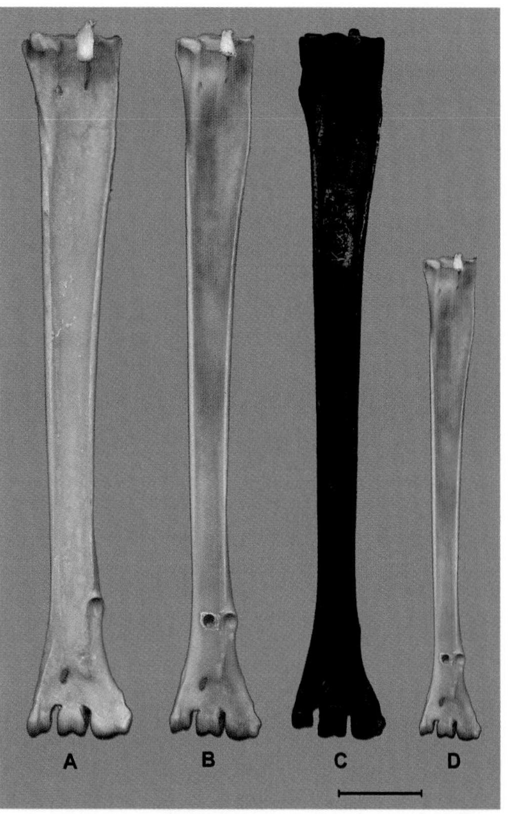


FIG. 2. Left tarsometatarsi of *Buteogallus* in posterior view: (A) *B. urubitinga* USNM 345786, (B) *B. meridionalis* USNM 630248, (C) *B. daggetti* LACM K3159, and (D) *B. meridionalis* USNM 630248. (A) and (B) are enlarged to the same size as (C). Scale = 2 cm for (C) and (D).

TABLE 1. Mean measurements (mm) (range in parentheses) of Savanna Hawk (*B. meridionalis*; $n = 7$) compared with those of *Buteogallus daggetti*. Measurements of the referred tarsometatarsus examined are essentially identical to those given by Miller (1915) for the holotype of *B. daggetti*; those for the tibiotarsus are from LACM J9744.

| | *B. meridionalis* | *B. daggetti* |
|---|---|---|
| **Tarsometatarsus** | | |
| Length | 106.6 (101.5–115.1) | 167.0 |
| Proximal width | 12.6 (12.0–13.4) | 20.8 |
| Least shaft width | 5.4 (4.9–6.0) | 9.1 |
| Distal width | 14.2 (13.3–15.2) | 22.8 |
| **Tibiotarsus** | | |
| Length from distal end of fibular crest to external condyle | 87.4 (83.4–92.6) | 142.6 |
| Least shaft width below fibular insertion | 6.6 (6.0–7.1) | 13.0 |
| Distal width | 12.5 (11.8–13.1) | 21.4 |

Fig. 3. Left tibiotarsi in anterior view: (A) *Buteogallus urubitinga* USNM 345786, (B) *B. meridionalis* USNM 630248, (C) *B. daggetti* LACM K3159, and (D) *B. meridionalis* USNM 630248. (A) and (B) are enlarged to the same size as (C). Scale = 2 cm for (C) and (D).

generic value, so *Wetmoregyps* must now be subsumed in *Buteogallus* when that genus includes *Heterospizias*.

## Discussion

*Buteogallus daggetti* was a rare bird even at Rancho La Brea and has not yet been recorded outside of southern California and northern Mexico. We may infer something of its probable habits by studying the habits of its most similar living relative, *B. meridionalis*. As its name implies, the Savanna Hawk inhabits open grasslands or savannas with scattered trees or shrubs. It may hunt from a perch but also forages on the ground, especially at the edges of advancing fires (Ferguson-Lees and Christie 2001). Prey is diverse, including mammals, reptiles, amphibians, and arthropods. During the wet season in the llanos of Venezuela, it feeds largely on crabs (Mader 1982).

The long, slender legs of *B. daggetti* indicate a significant terrestrial component in its behavior, similar to that of *B. meridionalis*, as Miller (1915) correctly inferred at the outset. Likewise, it also must have been an inhabitant of open country. Steadman et al. (1994:580) determined that grassland species made up a "very strong component" of the fossil avifauna of San Josecito Cave, though they did not infer habitat preferences for most of the extinct species, including *B. daggetti*. The puzzling report that Miller had later concluded "that *Wetmoregyps daggetti* was a forest inhabiting species rather than a walking eagle as previously considered" (Storer 1931: 177) turned out to be based on some spurious and somewhat circular reasoning. Because *Wetmoregyps* was more abundant in the much less numerous fossil material at Carpinteria than at Rancho La Brea, and because the environment at Rancho La Brea was believed to have been open, Miller (1931:369) concluded that the environment at Carpinteria must have been different so that "*Wetmoregyps* was a sylvan form." This assumption was based on only five bones of *Wetmoregyps* from Carpinteria, and Miller did not say how many individuals were represented. For a rare species, chance alone could produce 5 bones among a total of 1,000 at one site and 5 among 100,000 at another.

Miller (1928:255) was more accurate in characterizing the species as "weak-footed," so that one may visualize *B. daggetti* as a New World analog of the Old World Secretarybirds (*S. serpentarius*), which today are confined to Africa but which also inhabited Europe during the Tertiary (Mourer-Chauviré and Cheneval 1983). The two species of *Pelargopappus* in Europe had proportionately shorter tarsometatarsi than *Sagittarius*. Extrapolating from distal width, the length of the tarsometatarsus of the smaller of the two European species (*P. schlosseri*) would have been 168 mm, which is essentially the same as that of *B. daggetti*. Least diameter of the shaft of the tibiotarsus is a good indicator of body mass (Campbell and Marcus 1992). In captive specimens of male and female Secretarybirds weighing 2,685 and

3,330 g, respectively (USNM 621022, 621021), the least diameter of the shaft of the tibiotarsus was 31 mm, whereas in the specimen of *B. daggetti*, it was 35 mm. This indicates that this species was at least as hefty as a modern Secretarybird and, therefore, could well have occupied a similar niche. This is not unlike the situation in which the open country of North America produced felids convergently similar to Old World cheetahs (Van Valkenburgh et al. 1990).

Secretarybirds (*Sagittarius*), like Savanna Hawks, feed on diverse prey items but are best known for their predilection for snakes. *Buteogallus daggetti* may have had somewhat similar habits and would certainly have been capable of subduing much larger snakes than *B. meridionalis*. There appears to be no reason for considering that *B. daggetti* was anything other than an incidental scavenger, if that, contrary to the speculation of Steadman and Martin (1984). Thus, it is unlikely that the extinction of *B. daggetti* can be directly related to the extinction of the North American mammalian megafauna, unless some of the extinct mammals were responsible for the existence of suitable open habitats for the raptor.

## Acknowledgments

I am pleased to contribute to this volume honoring the memory of Ned K. Johnson, who was a supportive and appreciative colleague of mine for many years and who is sorely missed. Some of the ideas in this paper were developed in collaboration with S. Parry and W. Suárez. I am very grateful to K. E. Campbell, Natural History Museum of Los Angeles County (LACM), for making available fossil material from Rancho La Brea. I am grateful to S. Emslie and D. Steadman for comments on the manuscript. The figures are by B. Schmidt, Division of Birds, National Museum of Natural History, Smithsonian Institution (USNM).

## Literature Cited

Baumel, J., A. S. King, A. M. Lucas, J. E. Breazile, and H. E. Evans, Eds. 1979. Nomina Anatomica Avium: An Annotated Anatomical Dictionary of Birds. Academic Press, London.

Campbell, K. E., Jr., and L. Marcus. 1992. The relationships of hindlimb bone dimensions to body weight in birds. Pages 395–412 *in* Papers in Avian Paleontology Honoring Pierce Brodkorb (K. E. Campbell, Jr., Ed.). Natural History Museum of Los Angeles County, Science Series no. 36.

Ferguson-Lees, J., and D. A. Christie. 2001. Raptors of the World. Christopher Helm, London.

Howard, H. 1932. Eagles and eagle-like vultures of the Pleistocene of Rancho La Brea. Carnegie Institution of Washington Publication 429: 1–82.

Howard, H. 1936. Further studies upon the birds of the Pleistocene of Rancho La Brea. Condor 38:32–36.

Mader, W. J. 1982. Ecology and breeding habits of the Savanna Hawk in the llanos of Venezuela. Condor 84:261–271.

Miller, L. 1915. A walking eagle from Rancho La Brea. Condor 17:179–181.

Miller, L. 1925. The birds of Rancho La Brea. Carnegie Institution of Washington Publication 349:63–106.

Miller, L. 1928. Generic re-assignment of *Morphnus daggetti*. Condor 30:255–256.

Miller, L. 1931. Pleistocene birds from the Carpinteria asphalt of California. University of California Publications Bulletin of the Department of Geological Sciences 20:361–374.

Miller, L. 1943. The Pleistocene birds of San Josecito Cavern, Mexico. University of California Publications in Zoology 47:143–168.

Mourer-Chauviré, C., and J. Cheneval. 1983. Les Sagittariidae fossiles (Aves, Accipitriformes) de l'Oligocene des Phosphorites du Quercy et du Miocene Inférieur de Saint-Gerand-le-Puy. Geobios 13:803–811.

Steadman, D. W., and P. S. Martin. 1984. Extinction of birds in the late Pleistocene of North America. Pages 466–477 *in* Quaternary Extinctions. A Prehistoric Revolution (P. S. Martin and R. G. Klein, Eds.). University of Arizona Press, Tucson.

Steadman, D. W., J. Arroyo-Cabrales, E. Johnson, and A. F. Guzman. 1994. New information on the late Pleistocene birds from San Josecito Cave, Nuevo León, Mexico. Condor 96: 577–589.

Storer, T. I. 1931. Sixth Annual Meeting [Cooper Ornithological Society]. Condor 33:177–179.

Van Valkenburgh, B., F. Grady, and B. Kurtén. 1990. The Plio-Pleistocene cheetah-like cat *Miracinonyx inexpectatus* of North America. Journal of Vertebrate Paleontology 10:434–454.